The discovery of high-temperature superconductivity was hailed as a major scientific breakthrough, inducing an unprecedented wave of excitement and expectation among the scientific community and in the international press. This book sets this research breakthrough in context, and reconstructs the history of the discovery.

The authors analyze the emergence of this new research field and the way its development was shaped by scientists and science policy-makers. They also examine the various institutional and national settings in which the research was undertaken as well as considering the scientific backgrounds and motivations of researchers who entered the field following the original discovery. The industrial connection and the general belief in promises about potential future applications were important elements in strategies devised to obtain funding. A remarkable factor in this process was the role played by the media. The sustained attention that followed the discovery of high-temperature superconductivity resulted in it being seen as the symbol of a new technological frontier. This remarkable story and the developments that followed provide fascinating insight into the current changes that transform the system of science under the impact of international social and economic pressures.

This book will be of interest to scientists generally, to science policy-makers, to those interested in the sociology of science and technology, to students working on the public understanding of science, and to general readers interested in science and scientific development.

AFTER THE BREAKTHROUGH

AFTER THE BREAKTHROUGH

The emergence of high-temperature superconductivity as a research field

HELGA NOWOTNY
ULRIKE FELT

*Institute for Philosophy of Science
and Social Studies of Science,
University of Vienna*

CAMBRIDGE
UNIVERSITY PRESS

PUBLISHED BY THE PRESS SYNDICATE OF THE UNIVERSITY OF CAMBRIDGE
The Pitt Building, Trumpington Street, Cambridge CB2 1RP, United Kingdom

CAMBRIDGE UNIVERSITY PRESS
The Edingburgh Building, Cambridge CB2 2RU, United Kingdom
40 West 20th Street, New York, NY 10011-4211, USA
10 Stamford Road, Oakleigh, Melbourne 3166, Australia

First published 1997

Printed in the United Kingdom at the University Press, Cambridge

Typeset in 11/14 Times

A catalogue record for this book is available from the British Library

Library of Congress Cataloguing in Publication data
Nowotny, Helga
After the breakthrough: the emergence of high-temperature
superconductivity as a research field / Helga Nowotny, Ulrike Felt.
p. cm.
Includes bibliographical references and index.
ISBN 0 521 56124 8 (hardback)
1. High temperature superconductivity – Research – History.
I. Felt, Ulrike. II. Title.
QC611.98.H54N69 1997
537.6'23'072 – dc20 96–12444 CIP

ISBN 0 521 56124 8 hardback

Contents

Acknowledgments

Our study received support from many quarters. A grant from the Austrian Science Foundation (*Fonds zur Förderung der wissenschaftlichen Forschung*) allowed us to conduct the detailed study of three European countries that forms the empirical core of the book. Hans Chang, Director of FOM, the Dutch Foundation for Fundamental Research on Matter, invited us to add the Netherlands to our investigation of national research systems, although we had already finished our interviews in the other countries. The German Marshall Fund enabled us to visit one of the many well-attended HTS conferences, this one in Stanford in July, 1989. Many friends and colleagues in the natural sciences offered informal comments and their views on the promise of HTS.

A number of individuals provided us with indispensable and continuing advice, guidance, and assistance: Carlo Rizzuto, from the *Istituto Nazionale per la Fisica della Materia*, in Genoa, not only shared his vast professional experience in the field, but also provided greatly appreciated guidance in technical aspects and generous encouragement throughout our study. Ted Geballe at Stanford University, Rustum Roy at Penn State University, and Thomas Hughes at the University of Pennsylvania took the time to read early drafts of the manuscript and to give us invaluable comments. Magda Gronau, now at the *Wissenschaftszentrum* in Germany's state of North Rhine-West-phalia, compared in detail our empirical findings with her profound knowledge of the German science policy system. Jean-Bernard Weber, Secretary of Section II of the Swiss Research Council, and Helmut Rauch, President of the Austrian Science Foundation, were kind enough to check the accuracy of the sections pertaining to their respective countries policy-making. John Ziman encouraged us throughout the project and was extremely helpful in a decisive moment. Valerie Jones and in particular Mitch Cohen helped us transform some of our cruder Germanisms into more idiomatic English. Above all, we are indebted to the generosity of the scientists and science administrators who agreed to be interviewed, thus allowing us to follow the fascinating turns taken by HTS as a research field.

We especially thank the Collegium Budapest/Institute for Advanced Study for the generous and supportive environment it provided in the final phase of completing the book.

Events in our private and professional lives have prevented us from completing our study as rapidly as initially planned. We hope the unanticipated delay has its good side. Looking back on the development of the field at the end of 1995, nine years after everything began, provides, if not historical depth, then at least enough distance to place the extraordinary event in a broader context.

Budapest and Vienna
Helga Nowotny and Ulrike Felt

1

Introduction: the emergence of a new research field

Discovered in 1911 at temperatures near absolute zero, superconductivity is the loss of resistance to electrical current some materials display when cooled below a "critical temperature". The phenomenon was confined to scientific laboratories until the late 1950s, when first technological applications became feasible. It also took nearly half a century before a theoretical explanation of the phenomena – the BCS[1] theory – was formulated. In the following two decades, numerous researchers contributed to the field, but no materials were found with critical temperatures higher than 23 Kelvin ($-250°$ Celsius). By the mid-1980s, the scientific community had reached the consensus that superconductivity was a closed field, and that the dream of room-temperature superconductors should be abandoned.

But the year 1986 changed this situation dramatically. Two researchers at the International Business Machines (IBM) lab near Zurich, Switzerland, discovered a new class of materials among the ceramic oxides that display superconductivity at temperatures far higher than previously observed.

High-temperature superconductivity was born.

1.1 A surprising discovery and its consequences

Like a minor earthquake, the discovery of high-temperature superconductivity in late 1986 sent a shock wave through the research systems of the industrialized countries, exciting scientists, policy-makers, and the lay public alike. We followed the course of the discovery, intrigued to observe and analyze what the tremor revealed: which structures of the system of science and research

---

[1] The theory was developed by Bardeen, Cooper & Schrieffer.

proved robust and resistant, and what gave way, crumbling under the unexpected shake-up?

But above all, we wanted to investigate what researchers, science policy-makers, industry, the media – and through them the general public – would make of the event. We were interested in the now-bared interconnections between the different parts of the "building" – the strategic research hinges and organizational structures that connect basic research, technological applications, and worldwide economic competitiveness – and especially how and why thousands of researchers all over the globe rushed to exploit the new opportunities, how and why national funding agencies decided to support them, and how the media told the continuing story in a collective and collusive effort to establish a new field of research.

Comparing the unexpected discovery of high-temperature superconductivity (HTS for short) to an earthquake raises the question of the magnitude of the convulsion. In any case, its repercussions are lasting, since they follow some of the major fault lines of the current science and research system. Moving from a seismic to an engineering metaphor, we can compare the breakthrough to a "transient" or "impulse" load. Engineers deliberately induce a well-controlled stress to test the strengths and weaknesses of a system or structure. We approached the science and research system as if the discovery of HTS were such a stress. We focused on the responses of the various components of this system – researchers and research institutions, policy-makers and governments, industry and the media – to see how they bore up under the test, how well- or ill-prepared they were for unexpected events of this kind, and what their responses showed about the underlying processes, strengths, and weaknesses of the system as a whole.

We witnessed the emergence of a new research field in an extremely short time. Yet the participants' intense efforts to establish the field were at odds with the cool, detached view of the prospects so strongly claimed in public rhetoric. For us, the point is not whether the new field's achievements thus far – in theoretical or in practical (i.e. technological or commercial) terms – fulfilled the initial hopes and expectations, not whether funding was adequate, and not whether the behavior of scientists and the media was disproportionate, a transgression of the conventional norms of science. We see the field of HTS as typical of the present mode of scientific research, an extended mode that combines separate scientific disciplines and research fields and in which even basic research must reach out into society and thus into unknown but anticipated contexts of potential technological applications. This mode of research presupposes the active and strategic participation of a widened set of social actors who work in more – and more various – sites of knowledge production and in novel patterns of alliances and cooperative or competitive behavior. Researchers intent on remaining active players in the game must now involve themselves much more directly in preparing, shaping, and managing the conditions that enable them to pursue the kind of research they desire. The discovery of HTS was unplanned, but it has proven a good test case for the ability and preparedness of individual researchers, local research groups, and national research systems to organize the prerequisites of research.

HTS was one of those bursts of scientific innovation, based on a discovery unanimously termed a breakthrough, that every working scientist, science administrator, or policy-maker secretly or openly yearns for. Yet when it occurred, it took everyone by surprise.

Indeed, few scientific discoveries in recent years have triggered such a wave of excitement and acceptance among thousands of researchers around the globe as did the discovery of HTS. With the exception of the ostensible discovery of cold fusion, none has comparably fired the layman's imagination. The common faith – as it must be termed – was that new technological goals could be realized, if not in the short term then in the medium and long term. Hope arose for an era of renewed economic growth and change in people's lives comparable to those resulting from the semiconductor revolution. Visions of technological utopias revived.

For a brief moment, scientists seemed on the verge of even more exciting discoveries. One of the leading researchers in this field recently told of a recurrent dream he had when he first began work on high-temperature superconductors. He saw himself:

standing at the threshold of a vast, dark room, with the door open barely a crack, sending a narrow shaft of light onto the floor. I would try to open the door further to illuminate the whole room, but it would not move: I could only put my head inside and see what was directly lit, not what was hidden in the darkness. The feeling was one of longing, fear and expectation, for I desperately wanted to see what was inside, but I was afraid of what it might be and how I might respond. My subconscious was clearly grappling with what many were probably feeling at the time – that we were on the verge of a great moment – and that we would be part of, or at least direct observers of, a historic time in solid state physics and materials science

(Cava, 1993: 297).

But the private worlds of dreams and the shared hopes and expectations on which they were based were inevitably intertwined with the more sober public world, where funding for research must be obtained and where government policies, international rivalries, and economic competition prevail. Nine years after the discovery, the initial excitement has subsided and the work of those remaining in the field has long since assumed a more normal pace and pattern. National science and technology policies have been established to strengthen the national scientific base and thus put national industries in position to partake in any spurt of technological growth. Investment in basic research continues at a modest level. After an initial wait-and-see period, some of the larger industrial companies decided to pursue HTS research on a longer-term basis. Many small companies could not wait long enough to see their investments mature and had to withdraw.

Scientific journals still report the latest experimental results on the many remaining theoretical and empirical puzzles. New HTS compounds continue to be discovered, the most prominent recent example being the buckminster-fullerenes. While they differ fundamentally from the ceramic oxide HTS materials, they also show transition temperatures far higher than earlier limits. Their structure and composition is another surprising example of how atoms can fill space and of their resulting unexpected behavior. But no real advance has been made in understanding the physics underlying the superconductivity of such diverse substances as cuprates among the ceramics and the almost pure carbon of the fullerenes.

Although media excitement has subsided, hope is sustained by the knowledge that any technological innovation takes time. Cautiously optimistic reports on the production of technological devices still trickle in, such as Superconducting Quantum Interference Devices (SQUIDs), which incorporate HTS, and which are utilized in outer space and in medical diagnostics. But the general consensus is that it is still too early to say where HTS technologies will ultimately make their greatest contribution. Many feasibility studies predicted HTS would find a prime field in the microelectronics industry, but here economic competition is fierce, making it difficult for any new component to replace existing ones. The second realm of application commonly associated with HTS, the transmission of electric power, may hold surprises as well. Projections exist that the next growth phase of the electric system will be an order of magnitude larger than the one that has now reached saturation. Energy lost in transmission is a stable 10% – huge in absolute terms, but still judged too small to justify replacing existing transmission lines with HTS lines – however efficient they would be – in a continental or even intercontinental system. But superconductivity could revolutionarily reduce the size of machines, thus permitting the construction of units of larger capacity (Ausubel & Marchetti, 1993: 8–9).

HTS has lost its prominence in the media. After vociferously pushing the HTS story, especially in the United States, the media unsurprisingly moved on to newer issues, leaving the impression that the early hopes had been dashed and the future of the field looked less bright than expected. The many commercial HTS newsletters set up after the breakthrough are also gradually disappearing, except where they are mellowing into more conventional industry journals. The initial flurry of books on HTS, written largely by science journalists, has been followed by a second wave that places the discovery in a larger historical context (Vidali, 1992; Ott, 1992).

1.2 The extended laboratory and its constraints

The research field emerging in the nine years since HTS was discovered provides an example of the general trend of science and technology to move closer together. In this sense, HTS is an example of technoscience extending in whatever directions appear technologically promising. This brings us to the concept of the extended lab, a term used to characterize the vast and heterogeneous network linking each laboratory to its economic, political, administrative, technological, and scientific environment. Each network includes many partners who shape and define the content of research, the orientation of the programs, and the evaluation of results. Networks include not only individuals, but also resources, documentation, instruments, and funding. In this sense, a laboratory is not sharply separated from other production units in society. This is not surprising, since all competence and scientific knowledge has to find or construct its own space to circulate its products, if it is to sustain itself. Scientific research must continually create new products and generate demand for them (Callon, 1989: 13).

We agree with Callon that the modern technoscience lab extends into the wider society, rendering the boundaries increasingly diffuse, and that networks can be conceptualized as abolishing these boundaries altogether. But, however fluidly, the extended lab is structured, and thus subject to constraints. The activities of doing research, organizing the research environment, and projecting technological advances – activities constantly in flux and dependent upon previous outcomes – combine to give different organizational forms to the extended lab and to determine the primary directions of its extension. The first explorations of a new field are conducted without knowing where the most promising results will be found, nor how they can be turned into a technological advantage, taking cost, reliability, and other performance criteria into account. Thus, search strategies include strong random components, and it is the task of research organizations and their management to optimize this process (Montroll & Shuler, 1979). The degree of preparedness to grasp new opportunities varies markedly between research groups, industries, and countries.

Research is conducted on an international scale, so the extended lab has a geographic dimension. Researchers scattered around the globe in university, government, or industry laboratories are linked to each other through their work on related problems. Attending conferences, reading each others' publications, and communicating, they stimulate each other to further exploration. They converge in their realm of inquiry, which they populate in different modes (Becher, 1989). Exciting new questions may arise near those already posed. Obstacles may lead to temporary or long-term abandonment of a topic. Thus, the extended lab and the free flow of exploration are constrained.

The social organization of research is also a powerful constraint. Although many similarities exist, researchers in a university group conduct inquiry differently than those in a government or industry lab Management in industry is more likely to set deadlines for programs with make-or-break evaluations of the progress made. Industry management houses researchers together and acquaints them with company goals, thus providing more specific directions, even in basic research, than university researchers must deal with. But university researchers must obtain funding from research councils, which also impose conditions.

The organizational form that arose to coordinate the inquiry opened by the discovery of HTS was the national research program. Its intent was – without inhibiting the free international flow of information – to strengthen cooperation between researchers within a country, especially between university and industry labs., to increase the likelihood that the new discovery would yield tangible technological results. We will describe in detail the obstacles encountered and how university researchers responded to such attempts at management. But the point is that the extended lab is indeed constrained and structured by boundaries: those of the national research programs.

Another constraint in the extended lab is often overlooked. Any new product or device has to fit into the existing technological system, infrastructure, or production process. Technological development is a sequence of replacements in which cost and other performance criteria are limiting factors throughout. It may take a long time before the exciting "bright spots" that appear in scientific inquiry mature into technological exploitation. The history of technology shows that every innovation must pass through selective filters and constraints; these are often unforeseeable and contingent upon interaction with economic, political, and social factors. Initial conditions often determine the trajectories of technological developments (Arthur, 1988; David, 1985). Technologies may also become "locked in", evolving along their initial pathways, while changing course seems too expensive, even if alternatives are found that would have been more advantageous if they had been implemented first.

Though technological systems may appear impervious, Thomas Hughes has pointed out that they do not become autonomous, even after prolonged growth and consolidation. They acquire momentum, have a mass of technical and organizational components, possess directions or goals, and display a rate of growth suggesting velocity, but they must be maintained by the interests of people and organizations committed to them. Their robustness is partly that of technical artifacts and partly that of interests and institutions (Hughes, 1983). A case in point is low-temperature superconductivity; it is still too early to tell how, in the extended lab, with its organizational structure, and search strategies for technological innovation, HTS will cope with the competition of this already existing technology.

At this point, we would like to underline our own methodological constraints. Following scientists around in an extended lab differs from the method of studying them in one localized lab at a given time, as has been the case in most laboratory studies published so far (Latour, 1987; Latour & Woolgar, 1986; Knorr-Cetina, 1984). The range and depth of observations is necessarily less complete. Observations are limited geographically and temporally; they can be neither continuous nor simultaneous. Observations of a social process are based on inferences. They are always obtrusive, altering their objects. In the social studies of science, we do not speak of measurements. Instead, we attempt to listen to the various stories told by the actors we include. We observe and interpret how they construct their accounts. We weave our actors' contexts and the influence of these contexts on the construction of their narratives into the stories as tightly as possible. We confront the accounts with each other. In the end, we too can only tell our own interpreting story.

Moving around with researchers, science administrators, heads of research councils, and research directors of industrial labs in the extended HTS research lab thus provided the empirical basis for our account of the emergence of the new research field in the European countries we selected. The direct contact, interviews, and observation were necessarily limited in space and time. To follow the unfolding of the wider global context, we also utilized other sources: analyzing the scientific literature and of the general press; personally participating in international conferences; scrutinizing feasibility studies and other reports made to guide policy-making; and speaking with as many actors as possible. The overriding interest of our analysis remained the question: What does the case of HTS reveal about the present state of the science system?

1.3 Our study

Our analysis is based on more than 70 interviews carried out in 1988–89 with university researchers, science administrators in ministries, and representatives of various funding bodies and industry. Our interest focused on the gestation period following the discovery of HTS. Who joined the field, what were their motives, how did their different scientific backgrounds and skills shape the new research area, who pushed to set up national research programs and who were their allies, and what places became centers for HTS research? We were also interested in the speed with which and the extent to which national science policies adapted to a situation defined as exceptional.

In Austria, we interviewed the leaders of the research teams involved in the national coordinated HTS effort (*Stimulationsprogramm*), the leader of an independent team working at the *Atominstitut*, and representatives of the two industrial enterprises interested in the field, Elin Union AG and Metallwerk Plansee AG. We attended some of the national network meetings and two organized by the funding agency: an early one to enhance university–industry cooperation and a later one to evaluate the first two years of the national HTS program. We rounded out the information with discussions with representatives of the program's funding agency (the Austrian Science Foundation, FWF), who also gave us access to their written records.

In Switzerland, we visited the major groups involved in the national HTS program (SUPRA2), the national funding agency (Swiss National Science Foundation, SNF), and two industrial enterprises, the multinational company Asea Brown Boveri (ABB) and Spectrospin, a small company specialized in conventional superconductors. Regrettably, due to the "mainly confidential nature" of the subject matter, International Business Machines (IBM) did not grant us extensive interviews, which would have been invaluable in studying the influence exerted by a multinational research organization.

The German case was more complex, because we had to make choices on several levels: the regional distribution of the groups; the different forms of funding chosen; the variety of organizational structures; and the balance between basic and applied research institutions. Of the 15 associations of university groups set up in early 1988, we chose Tübingen, Cologne, Aachen, and Göttingen[2]; we also visited the Max Planck Institute in Stuttgart and the two "big science" laboratories in Karlsruhe and Julich. We took a closer look at the responses of two industrial companies, Hoechst and Siemens, and interviewed representatives of the Federal Ministry and the program managing

---

[2] In making our decision, we were advised by the coordinator of the program management agency, Magda Gronau.

agency. To a degree, these choices were arbitrary. The strength of a research group in the very early phase did not automatically mean it would continue in the future, nor did latecomers necessarily play a subordinate role. But our main interest was in the formation process of the new research field.

The interviews in Austria, Switzerland, and Germany were carried out in 1988–89, when HTS research was at a relatively early stage. From a micro-level perspective, we observed the hopes triggered by the emergence of a new field and how funding agencies and industry responded. During a brief first phase, funding was relatively open. In the early 1990s, most countries entered a second phase, in which funding became more selective and focused. At this stage, we were invited to extend our analysis to the Netherlands, giving us a second empirical glimpse of a later stage in the development of HTS research and in the making of national research programs. The number of actors shaping the Dutch national research policy was small enough to allow the emergence of a coherent picture of the sequence of events and of the factors contributing to the outcome.

The interviews were semi-structured and almost always conducted within the respective person's institutional context. With a few exceptions, our interview partners were the team leaders or senior researchers; this undoubtedly biases the sample. We also asked them to fill out a questionnaire, but it revealed little. We were thus provided with a comparative overview of the scientists' actions, their motivations, and the difficulties various groups involved in HTS faced in the three countries. To test some of the impressions we gained from the interviews and to round out the picture, we visited some university labs for a longer period of time. We hoped for a better understanding of specific characteristics of the national science systems: their historical evolution, personnel structures, funding mechanisms, and other local contingencies.

In contrast to the prolonged, in-depth study of a single lab, as usually found in the social studies of science, we faced a sample of university institutes, research labs, and industrial enterprises with a broad geographical distribution, great historical complexity, and wide organizational diversity. A long-term, comparative, micro-level observation of scientific activity was clearly unfeasible for us. Forced to modesty, in the end we chose to visit at least one German, one Swiss, and one Austrian university lab. For one week in each lab, we recorded our impressions of working conditions, the everyday problems researchers face, the way routine research was conducted, the communication mechanisms within and beyond the laboratory, and the influence of all these factors on the research done. Such relatively short visits amounted to snapshots of the researchers present that particular week and of the parts of the experimental program we were shown. Thus our description does not pretend to be the full picture, and even the balance among the research topics may have been distorted by how our discussion partners presented their worksite to us. But we did gather a wealth of valuable information and data.

In Austria, we chose to visit the Technical University of Graz, home of the largest association of research groups taking part in the country's coordinated study of HTS. It was one of the first groups to apply for joint local project funding. We knew from their publications that they were also active on the international level. In Switzerland, we visited the University of Geneva, which has a long, internationally highly-reputed tradition in the field of superconductivity and where four professors and their teams were working on a joint project on HTS. In Germany, again, the choice was more complex, due to the large number and organizational variety of institutions involved. In the end, we chose the University of Cologne, where a special research program (*Sonderforschungsbereich*) on HTS had been set up, thus promising an interesting additional organizational feature not found in other German labs.

We made our visits between November 1989 and January 1990. HTS had entered a phase of consolidation, and permanent, more focused programs had begun to replace the first ad hoc initiatives.

To gain a more global perspective, we participated in three international conferences, in 1988 in Mauterndorf and Interlaken and in Stanford in 1989. Further, we followed developments in the United States, United Kingdom, and partly Japan on the basis of gray literature and accounts written by other scientists.

1.4 Scope and outline of this book

Students of social studies of science and technology may now want to look back to ask what the excitement was all about and what is still being done, assessing general and specific features of the development of HTS. This book seeks to retrace the achievements of the first phase of intellectual excitement, intense scientific work, media overselling, and the funding agencies' cautious but sometimes misguided policy-making.

Our interest is not in judging whether initial expectations were too high, whether investment in research was justified, whether expenditures should have been greater or less, or what the many technology assessment and feasibility studies should have considered. Rather, we remain intrigued by the degree to which the discovery of HTS gripped the scientific and lay imagination. This excitement was transformed into more tangible and long-lasting results: the emergence of a new research field and its organizational framework in national research programs. Were the features of this development unique to HTS, or were they indicative of processes typical for the current research system? An answer to this question must embrace several perspectives.

First, there is the historical perspective. Although HTS is a novel phenomena, it has a fascinating specific prehistory. In the past, superconductivity has attracted some of the best minds in physics, and the new aspect of HTS continues to excite a large number of talented theorists, experimentalists, and engineers from a wide variety of fields. The manifestation of superconductivity in artificially structured materials made it one of the most challenging research subjects in materials science, which attempts to make or discover new materials, determine their physical properties, grasp the reasons for their behavior, and finally to find applications for them. We investigate which specific materials science research modes have been embodied in HTS research. Then we place HTS in the context of the development of applications in the field of low-temperature superconductivity, which preceded it. Next, we address the event that shook the physics community and beyond: the actual discovery of HTS by Müller and Bednorz at the IBM laboratory in Zurich. We maintain that the discovery was unlikely, and investigate the features of the research system that allowed it to occur nevertheless. We then trace the frenetic, euphoric phase that followed the breakthrough. Eventually, research patterns returned to normal; by then the foundations of a new research field had emerged.

The second perspective is that of the main actors and the knowledge they brought to the field: the scientists who were drawn to the new discovery and decided to work in the new field. We investigate the reconfiguration of persons from varying institutional and disciplinary backgrounds and their motivation to join the swelling numbers of those who produced some 18 000 publications within a mere four years of the HTS discovery. We examine the relationship between the newcomers and researchers who had worked in the "old" field of conventional superconductivity, or LTS, and the role that the construction of scientific expertise played in the making of the new research field. The exciting prospects presented an opportunity and the necessity to actively shape the research environment: scientists had to exert effort devising strategies to obtain funding. In the process, they had to forge various kinds of alliances; they had to make promises and set up structures that made these promises credible. They had to mobilize resources of various kinds not associated with the conventional image of scientists' behavior. In shaping the research environment and setting up national programs, they displayed how the cognitive and social aspects of scientific activity interact inseparably in communication, competition, and cooperation. We investigate the institutional and other factors influencing the observed outcome. Finally, we delve into the scientists' perspectives with a personal glimpse of everyday life and working routines in some of the labs we visited.

The third perspective is usually subsumed under the term science policy. Science policy is not a set of explicit statements of goals or of measures intended to achieve goals. It is fluid in concept and practice. It is structured around large institutional complexes of university, industry, and national research – but these are far from homogeneous, varying greatly from country to country and within individual countries, even over short periods of time. Individuals play a sometimes decisive role, but they must take into account institutional constraints and the actions of others, as well as having good timing in grasping opportunities. The specific form science policy usually takes is that of a national research program.

In the case of HTS, such programs were established with the urgency of the international competitiveness that was constantly invoked. But HTS was also seen as an attractive and highly visible test case for confirming or altering whatever science policy was in place or under discussion. HTS marked a turning point, but underlying structures and institutional relations in the science system proved remarkably stable. We analyze the conditions under which decisions were made to set up and fund national HTS research programs. We note contrasts between the responses of large and small countries to the worldwide challenge.

A remarkable phenomena was the new role played by the media, whose immediate and sustained attention allowed HTS to be seen as the symbol of a new technological frontier and a key to maintaining or advancing economic competitiveness in the international context. If the media had not reported the claims of ever-higher critical temperatures, scientists would not have been tempted to engage in "science by press conference". We examine coverage of HTS in some major American and European newspapers and analyze the influence of national cultures on the media construction of the "HTS story". We attempt to assess how the media's altered relationship to science affected scientists' communication patterns.

Our final chapter draws together the various strands while laying bare the gaps separating rhetoric and practice, beliefs and reality. We return to our original question: What is unique about the discovery of HTS and the developments after the breakthrough, and what do they indicate about the changes taking place in the research system? In the process, we re-examine some of the tacit and explicit assumptions that guided research in its "golden age" and the presumably irreversible changes in the research culture occurring now. The case of HTS clearly shows that these changes cannot be understood by looking solely toward the institutional, economic, or policy aspects, nor toward the unfolding of new scientific discoveries. Scientific creativity is inseparably linked to the innovation machinery of the present research system, which sustains it in a self-organizing capacity. The anticipation of potential technological applications has invaded basic research. The culture of science and the work of scientists is changing rapidly under the impact of social and economic complexities.

2

The context of the discovery

The discovery of materials that become superconducting at temperatures higher than previously observed or thought possible opened up a new research field. This chapter examines the individual, scientific, and institutional background of the discovery by Georg Müller and Alex Bednorz at IBM's Rüschlikon Laboratory in Zurich, Switzerland. The first section places the discovery in the context of the evolution, organization, and salient characteristics of the multidisciplinary field of materials science. Section 2.2 examines the industrial connection in an earlier period of technological optimism. We compare current hopes and efforts connected with the technological potential of HTS with the bright outlook for conventional or low-temperature superconductivity (LTS) in the 1960s and early 1970s. Few LTS applications materialized and only one proved commercially viable. What were the main reasons for the decline of the LTS field?

The third section presents a brief historical account of the study of conventional superconductivity and analyzes some of the factors that contributed to the new discovery, which was unexpected in terms of the discoverers themselves, the site, and the conventional wisdom refuted. Section 2.4 deals with the scientific community's reactions to the Zurich discovery. This highly unusual and intense period engendered some unconventional behavior in participants. Scientific excitement was flanked by passionate accounts in the media, which fueled public expectations about the technological and commercial significance of the breakthrough. Finally we describe the inevitable cooling-down phase that prepared the way for the establishment of national research programs.

2.1 Materials science as a research field

Individual scientists dominate the story of the discovery of HTS, but the initial event took place in the scientific and technological context of the field of materials science. Materials science spans a wide range of topics, including organic materials, metals, surfaces and interfaces, phase transitions, and new and artificially-structured materials. The latter constitute a new class of materials with intentionally produced spatial variations. Among them are the semiconductors, which have found the largest number of techological applications, as well as materials with novel electronic and magnetic properties. When a new compound has been found or new properties measured, the next step is to develop ideas about other compounds with similar structures, which leads to broader measurements and in turn to the synthesis of other new materials. The discovery or synthesis of new compounds is thus a step leading to, but not identical to, the analysis and measurement of the properties the compounds can be induced to display. This has resulted in a kind of international division of labor regarding these related but distinct activities.

Materials science research characteristically leads in a number of directions, thus opening up a variety of application contexts (Psaras & Langford, 1987). A relatively young field dating from the 1970s, it is uniquely suited to illustrate what is meant by an extended lab Difficult to define, it typically cuts across a spectrum of traditional disciplines to include metallurgists, physicists, chemists, crystallographers, engineers, and others. It is a sprawling realm whose institutional structure varies greatly from country to country in accordance with different scientific traditions and the contingencies of institutionalization. Precisely the diversity of the contributing disciplines stands in the way of strict internal cohesion and the integration of the field as a whole. The occasional semblance of disorganization and the turbulence of materials science distinguishes it from older, more tightly organized fields and reflects its cognitive content: its paradigmatic structure allows a variety of disciplinary approaches, instrumentation, and techniques, and it has developed on several research fronts simultaneously. It is a complex system devoted to studying material systems of increasing complexity at the microscopic level by combining increasing numbers of elements from the periodic table in controlled ratios and forms. For example, progressing from systems with two elements to those with three or four resulted in the discovery of supermagnetic materials and high-temperature superconductors. A practically infinite number of new materials can be produced and made to perform a wide range of mechanical, thermal, electrical, magnetic, acoustic, or optical functions. In principle and aim, materials are created "to customer order", designed for specific functions and properties.

The trend in materials science and in modern technology as a whole is a shift from brute force at the macro-level to information-rich synthesis at the micro-level. Multiple and flexible linkage between conceptual approaches, observational instruments, and the ability to control and manipulate the phenomena under investigation is the basis for a wide spectrum of applications (Rizzuto, 1991). New materials are found for specific functions, for example with much lower energy dissipation and investment for microelectronics. Superconducting materials can also be seen in this context: instead of increasing voltage to overcome resistance, electrons are persuaded to move in a non-dissipative way.

The paradigmatic structure of a research field implies that its communities of practitioners resemble each other in the way they approach and identify problems and in their judgment of what constitutes a solution. From the outset, HTS researchers had to cross disciplinary boundaries, find collaborators, share equipment, and borrow from the available variety of techniques and methods. Whether dealing with new and artificially-structured electronic and magnetic materials, adopting new ways of viewing surfaces, improving materials synthesis and processing techniques, or working with thin films or bulk materials, they were entering territory newly opened by the Zurich discovery at the same time as they contributed their own skills and knowledge. Thus, while HTS posed exciting new questions, the prevailing research modes and observational instruments were already in place. The initial vagueness about content displayed in funding proposals seeking support for HTS research only partly reflected the newness of the field; such vagueness is inherent in the research mode of materials science in general.

The discovery itself also conformed to a pattern in materials science. The researchers at Rüschlikon produced a new compound and measured its properties, a primary method of discovering new physical and chemical phenomena and of learning how the atomic, electronic, and bulk structures of materials lead to their observed properties. A diverse and heterogeneous field, materials science encourages the use of a wide variety of techniques and their associated instrumentation to deal with the complex, multi-faceted problems it addresses. From the beginning, some of the frontiers of materials science have required expensive facilities (for example neutron facilities in the 1950s). This led to attempts to coordinate an otherwise highly diversified field. In the United States, lab facilities were centralized to share the burden of costs and maximize access to equipment that few laboratories could afford on their own. In many countries, the discovery of HTS reiterated the questions of how well equipped research groups were, especially those at universities, and of what could be done to ease access to existing equipment.

Despite the use of large facilities, the most common form of research organization in materials science remains the small group, whether in university, industry, or government labs. At universities, the small research team is also the basic unit of graduate education and training, so the quality of research performed here is crucially dependent on the quality of student training. For major discoveries in materials science, small group research has a good track record (Di Salvo, 1987; Johnson, 1987). And HTS is a prime example.

Materials science's relatively loose organizational structure has resulted in widely dispersed research efforts. It may seem obvious that interdisciplinary research requires the coordinated efforts of several groups, but the record of success here is poor. One pioneering effort to set up a coordinating structure was the creation of the Materials Research Laboratories in 1959–60 in the United States. The Soviet Sputnik launch stimulated the belief that America's future security depended on maintaining a technological edge, especially in materials. Forty-five university centers submitted proposals to provide the interdisciplinary nexus deemed suited to ensure rapid advances in the field. The Department of Defense Advanced Research Program Agency (DARPA) chose three (Sproull, 1987). The component disciplines included chemistry, physics, metallurgy, mathematics, geology, and various branches of engineering. The envisioned applications ranged from consumer products to national defense.

Work was organized under an umbrella contract providing long-term government support in the form of equipment and facilities. The longevity of commitment allowed universities to allocate space on campus to materials research, to promote informal contacts among disciplines, and to provide central experimental facilities with advanced equipment, such as for electron microscopy and crystal growth. The creation of materials research centers also encouraged interaction between younger and older researchers, electrical engineers and chemists, and administrators and bench scientists. Job opportunities opened up for PhDs trained in the program. In 1961–62, the number of materials science labs grew from the initial three to twelve (Schwartz, 1987). When the US National Science Foundation (NSF) took over the program in 1972, materials science had become a recognized interdisciplinary field at many major US research universities. Academic departments changed their names, substituting the term "materials" for "minerals and mining" or "metallurgy" (Schwartz, 1987: 37). The educational environment took a radically different path than in the past. Where students had once associated almost exclusively with advisors and peers within their field, now new collaborative efforts and unforeseen results were possible.

In Europe, no such concerted effort was staged. But discoveries of new phenomena or of properties of new materials have continued over the last twenty years. Usually they have been made on the margins of research directed toward other, often unspecified objectives. Many new compounds were found by accident when researchers systematically examined phases in materials that resulted from new combinations of elements.

The list of major discoveries in materials science prior to HTS exhibits an interesting pattern: European research groups tend to discover new compounds, while US groups observe new phenomena (Di Salvo, 1987: 163). This international division of labor reflects the differences between a chemistry- and a physics-based approach. Europe's record in organic and inorganic chemistry, especially solid state chemistry, is better than the United States'. After years of watching their US colleagues exploit their discoveries, European researchers began reorganizing and broadening their range of interests to include physics. The shift was most pronounced in France, where research groups within the Centre National de Recherche Scientifique (CNRS) and elsewhere no longer restricted themselves to publishing the results of systematic studies of the synthesis and structures of new compounds, but also dealt with their electrical, magnetic, thermal, and other properties. The work of the French group that provided an important clue for Müller and Bednorz is part of this tradition exemplifying a strategy combining US and European approaches.

Over the years, materials science has moved from the laboratory into industrial production processes. Rather than starting with a material and developing a product to exploit its traits, now materials are often sought to fulfill the specifications of an envisioned product (Cohendet et al., 1987). The real novelty is not so much in the materials (where one can speak simply of improvement) but in the altered relationships between materials, the process of production, and the product. The value of so-called "new materials" is still only about 5% of all materials sold, but the competitive pressure of advances in materials science can be expected to force older and traditional materials (such as steel, wood, glass, plastics, etc.) to respond with improvements of their own, whether in purity control, computer design techniques, or production methods. Innovative combinations of old and new may also be in the offing.

Materials science has advanced by improving methods of understanding, analyzing, and manipulating matter at the microscopic level. While the old production technologies deformed bulk matter at the macroscopic level and standardized processes and products, the new production technologies operate at the microscopic level; at the extreme, it is now possible to relocate individual atoms. This opens the door for a remarkable change. Rather than standardizing the product, we are moving to a situation in which material can be produced "to order", varying with the functions or performance desired.

Thus, the general trend exemplified by materials science increasingly applies to production technologies. It is a shift from reductionism to complexity, from brute force to a non-intrusive, synthetic approach. It is based on increased performance in three areas: observational instruments (which are also moving out of the lab into industry); mathematical and computational instruments; and the manipulation and control of the microlevel of materials. Research now experiences more synergy and exchange between different disciplines sharing new concepts or instruments. More research is being done in directions not fixed in advance.

2.2 Technological optimism revisited

The long road to applications in superconductivity

The history of superconductivity is fascinating. Many great scientists, from Einstein and Bohr to the London brothers, from Landau to Feynman and others, had sought a deeper theoretical understanding of the phenomenon, and many illustrious experimentalists had also studied it. Superconductivity has inspired dreams of technological applications ever since Kamerlingh Onnes discovered it in 1911 (Kamerlingh Onnes, 1967). The resistance-free conduction of electric current and the repulsion or trapping of magnetic fields are macroscopic manifestations of quantum-mechanical effects normally observable only at the atomic level. Under certain conditions, these properties give superconductors great advantages over other materials. Superconducting generators, for example, have power to volume ratios up to five times those of conventional generators; high-field magnets like those used in particle accelerators consume much less power. In addition to energy-efficient devices, superconductors are also used in sensors.

The discovery of HTS was preceded by seventy-five years of experience with "conventional", low-temperature superconductivity. Whatever future HTS applications were projected, they inevitably invited comparison with earlier, unfulfilled hopes for LTS. Researchers moving into the "hot" research area met with groups who "came in from the cold", with an entire generation of researchers' memories of hopes, frustrations, and achievements.

The development of scientific theories has often been compared to a maze in which some paths have been explored in vain, while others have led to success. Superconductivity was no exception. Avenues abandoned in one area were sometimes taken up elsewhere (Pippard, 1977: 2). In hindsight, the labyrinth of scientific investigation often appears as if it were a straight path to the solution of problems. In 1986, the BCS theory was still believed generally valid, but the discovery of HTS brought several old, unsolved issues to the fore again. So there were good reasons to return to paths of inquiry that earlier generations had abandoned. LTS researchers, with their accumulated tacit theoretical and experimental knowledge, were essential to the new field.

For researchers who had taken part in the 25 years of LTS research, HTS offered the attraction of familiarity, though the differences were very great. Several LTS researchers held prominent positions and spoke with the authority of experience both for the importance of the new discovery and against being swept away by enthusiasm. Bardeen (the two-time Nobel prizewinner nicknamed the "gentle giant") and his colleagues Cooper and Schrieffer were welcome keynote speakers at the major conferences. Ted Geballe and John Hulm, two leading figures in the "old" search for LTS materials, reviewed the prospects of the new superconductors in a seminal article (Geballe & Hulm, 1988).

But this earlier generation of researchers had also witnessed the rise and fall of their field when the focus of work shifted to developing practical applications. They had watched while many of the large-scale LTS national programs established in the mid-1970s were phased out. Many had lost interest in superconductivity and turned to other fields or engineering.

LTS had originally flourished in the period of unbounded technological optimism of the 1960s. One reason for this optimism had been the plethora of funding in what was indeed a golden age for research. Any proposal with futuristic appeal stood a good chance of being funded; a realistic assessment of its potential was considered an asset, but by no means a necessity. Unlimited energy was expected from nuclear fusion and superconducting electrical networks. Space program administrators believed sufficient money and staff could put into orbit whatever the scientific imagination could dream up. Many countries established major national programs to utilize superconductivity in both the civilian and the military sector when it finally found a number of applications. The BCS theory was firmly in place: electrons were envisioned as traveling in pairs, "helping each other" bypass the resistance offered by atoms.

In 1961, a technological breakthrough was achieved with the discovery of niobium–titanium, the high-field superconductor used to make wires for high-field magnets, and with the understanding of flux pinning in current-carrying wires. At last, it seemed as if superconductivity's status would shift from that of a laboratory curiosity to an applied technology. But the interval separating lab advances from industrial utilization exceeded expectations, with many major breakdowns and failures of wound magnet coils.

Not until early 1970, after a series of improvements in reliability, was the first viable superconducting wire produced. The primary market for it was, and remains, other specialized scientific programs, such as particle accelerators, fusion experiments, and observation devices used in high-energy physics.

Magnetohydrodynamic conversion, magnetically levitated trains, power generation advances, and marine propulsion systems have so far never left the drawing board. The only successful commercial application of LTS wire to date has been in magnetic resonance imaging (MRI), developed in the mid-1970s. Superconducting magnets soon captured more than 95% of the MRI market. Compared to the expected success on the power and transmission market, this was still small change – about five hundred machines worth a billion US dollars a year in the United States, with General Electric and Germany's Siemens Corporation controlling more than half the world market.

Throughout the 1970s, technical reliability had greatly improved with the introduction of cryostatic stabilization techniques, and multi-filament niobium–titanium wires were now used in magnet production. The high expectations for LTS cannot be termed unreasonable or lacking in technical foundation. Hope centered on the adoption of superconducting components in large-scale energy production, storage, and transmission systems and in sensors for use in microelectronics.

A NATO Advanced Study Institute held in 1973 reviewed the achievements and prospects for large-scale applications of superconductivity and magnetism. In the US, the search for alternatives to foreign oil, which had jumped in price during the oil crisis, boosted interest in superconductivity. The variety of ideas for utilization was great: power generation and distribution; marine propulsion; magnets for commercial fusion reactors; high-speed ground transportation, such as maglev (magnetic levitated trains); industrial processing, such as magnets for separating impurities; sensors for process and quality control; and compact accelerators for the X-ray lithography of microelectronic chips (Foner & Schwartz, 1974). Prototypes of several of these applications were already available or at an advanced stage of development. The future looked bright.

Yet, the road towards further advances was – and still is – winding and arduous.

LTS as a system component: a contingent technology

Of the long list of uses for LTS that seemed feasible in the early 1970s, very few have been realized. Were engineers, investors, and government agencies merely indulging in technological dreams without considering the likely obstacles? Whence this discrepancy between promise and reality?

LTS is a minor but often essential component of other, larger technological systems, and its success has thus depended on the fate of those systems. The LTS discrepancy can be seen as an example of a kind of bottleneck in the genesis of large technical systems that was bound to occur sometime before the larger systems consolidated and developed their own momentum. Large technological systems almost never develop according to the projections of their designers (Joerges, 1988: 26–7). What determines the evolution of such systems and the integration of their components?

It soon became clear that the commercial success of LTS depended on a number of additional factors. First, solutions to problems of reliability and safety would have to be found, since a complex system is only as strong as its weakest component. Thomas Hughes' "reverse salients", i.e. system components that have fallen behind or are out of phase with the others, would have to be overcome. The failure of a superconducting device can result in the failure of the entire system; so the cores of such systems provide considerable resistance to the adaptation of such devices (Hughes, 1983).

Second, wherever superconductors were not the only way to do something, they would have to outperform or underbid conventional parts. Indeed, LTS had spurred a number of improvements in its more traditional competitors.

Third, as a component of larger systems, superconductivity would remain a contingent technology. For example, superconductors would be indispensable to a fusion power plant, but their use there still depends on the overall development and implementation of fusion.

Power transmission is another example of contingency: superconducting transmission lines promised higher efficiency than existing overhead or underground lines, which lose a considerable amount of the transmitted electric power due to resistance. In the 1970s, several studies were made of AC superconducting transmission lines. One, the Brookhaven project, met all its design objectives and maintained its efficiency over a lifetime of fourteen years. But by the time it was completed in 1986, US energy consumption was no longer expected to double every ten years. The short-term demand for new transmission lines of any kind had fallen together with the demand for new generating capacity. Utilities in the United States now prefer to build small power plants close to the end users, rather than large plants far away (OTA, 1990: 36).

Superconducting generators are a third example of dependence on larger systems. Several prototypes had been designed and constructed at Massachusetts Institute of Technology (MIT), General Electric, and Westinghouse in the United States in the 1960s and 1970s. But no full-scale generators were built, again due to existing surplus capacity and the expected fall in the demand for energy. In Germany, Siemens is still planning a commercial 850 MW system, and a consortium of Japanese firms is currently developing a 200 MW generator for the late 1990s (OTA, 1990: 33).

Another hurdle to the utilization of LTS is the low temperature it requires. To the complications and expense of cryogenic technology came the fear that

US helium wells were rapidly depleting, endangering the safe, long-term availability of the gas. But the discovery of HTS in 1986 meant that superconductivity no longer depended on helium.[1]

The contingency of superconducting applications is also highlighted by national differences or asynchronicities, often the result of political decisions. When enthusiasm for LTS began to ebb and programs were being abandoned in the US and Europe in the 1970s, Japan had just entered a phase of technological optimism and was able to set up and continue a number of industrial programs, reaping many of the benefits accruing to newcomers in the process of technological innovation and diffusion. Then as now, national programs were often instituted to enhance national prestige or industrial competitiveness (Isawa, 1974). Transportation systems like the maglev are notoriously dependent on political decisions and face enormous competition from air and auto transportation lobbies. In the United States, all support for high-speed ground transportation research ended in 1975, but the Japanese continue work on a superconducting maglev system based on the US magnetic levitation and propulsion scheme. Ironically, when the US was re-evaluating the advantages of high-speed ground transportation, the lack of domestic maglev technology meant it had to consider German and Japanese systems (OTA 1990: 38–9). Similarly, in 1983, IBM closed down its three hundred million dollar Josephson junction computer project in the US, just as Japan was taking up the idea. Whether Japan has successfully leapfrogged in either LTS or HTS remains to be seen.

Thus, superconducting applications have been partly dependent on government-supported technological systems, themselves at the mercy of extraneous political and economic circumstances. In the 1970s, superconductivity was expected to play a role in a context of rising energy demand and the attendant rapid expansion of the nuclear industry, the large-scale development of new transportation systems, and unabated economic growth. When these programs were scaled down, superconductivity was adversely affected by funding cuts as well as by the reduced prospects for the system as a whole. Some technical problems – like harnessing fusion – can't be solved by throwing money at them, but only by slowly accumulated expertise.

Jean-Claude Derian has likened technological dreams to a "mirage": highly desirable but ultimately unattainable (Derian, 1990). But even in hindsight, the

---

[1] Incidentally, the rapid development of cryogenics has shifted expectations for HTS. Between 1970 and 1990, cryogenic technology has achieved very high reliability. Helium availability has increased with the improvement of extraction techniques. Liquifiers, refrigerators, and cryostats developed for physical and medical applications are now so efficient and reliable that some HTS materials may be used – not because they have higher critical temperatures, but due to other good qualities – *alongside* LTS at liquid helium temperatures.

technological optimism for LTS does not seem completely unfounded. It did not ignore technical, economic, and political constraints, but assumed they could be overcome. The success of LTS as a contingent technology depended on the future of the systems and programs utilizing it – which looked bright at the time.

Magnetic resonance imaging (MRI): the successful exception

The most successful commercial application of LTS has been magnetic resonance imaging, MRI. Developed in the mid-1970s, it made the medical diagnostics industry the largest non-governmental market for superconducting wires and magnets. MRI is a powerful tool for diagnosing a variety of internal medical disorders. Costing some seven hundred US dollars per scan, it is more expensive than X-ray, computerised axial tomography (CAT)-scanning, or ultrasound imaging techniques, but its superior images has secured it a small but stable market (Hunt, 1989).

MRI had a chance to succeed because the market for medical diagnostics equipment is accustomed to the early adoption of specialized technologies. High performance, not low cost, is the purchase criteria. Moreover, the high cost is spread among the major care providers of the health system and mitigated anyway by the value of early, accurate diagnoses to both the patient and the health care system. The first MRI imager was built in 1978, so the first superconducting magnet system, introduced just two years later, didn't have to displace a well-established, well-optimized competing system.

Perhaps the most interesting aspect of MRI in our context is that it was unforeseen when superconducting wires were first developed twenty years ago. This has implications for HTS: many of our interviewees believe it is impossible to predict what future HTS applications will be the most important, but the first of them may well be in specialized markets where high performance is more important than cost considerations. Defense, health, and other research sectors are likely candidates.

A diffusion sequence for HTS

Based on historical precedent, economic rationale, and existing data on superconductivity, DeBresson has developed a general diffusion model predicting a new technology's most probable diffusion pathway between industrial sectors (DeBresson, 1991, 1995). Initially, new technologies need only demonstrate technical feasibility; cost factors are unimportant. Since demonstration of technical feasibility is costly, public funds are usually used, for example in government research labs, the military sector, space research, medicine, and scientific research itself. LTS was developed in this way.

Next, all technologies seem to follow a predictable path from instruments to machine tools to machines, then to industries that transform materials and primary goods, next to energy, transport and communications systems or utilities, and finally to durable consumer goods industries. As these industries adopt the new technology in sequence, the criteria of introduction are increasingly economic and less and less purely technical. By the time large energy distribution, transport, and communications systems – often regulated monopolies – adopt the technology, prior applications have resulted in some standardization to make components compatible. Standardization in turn permits mass production, economies of scale, and the reduction of unit cost. This is the threshold for the consumer durable sector, characterized by high price elasticity.

DeBresson argues that this sequence of diffusion crosses various thresholds of required peformance. The first peformance criteria, at the technical feasibility stage, is that a technology functions at all. At the instrument stage, performance must also be reliable. The machine tool sector further demands technical efficiency. So far, all criteria for adoption are primarily technical, not economic. In the machinery sector, economic criteria begin eclipsing strictly technical requirements: here, the technology must promise increased returns in the long run. Industries transforming materials then demand that the new technology be adaptable and versatile – and economically efficient enough to warrant replacing the currently used technology. Next in line, large and complex transport, energy, and communications systems demand system compatibility and economic reliability. Finally, the durable consumer sector requires user-friendliness, economies of scale (for mass consumer goods), or a new satisfaction (for luxury consumer goods).

What does this model predict as the most likely sequence for industry's introduction and acceptance of new HTS devices? The research, health, and defense sectors provide a market for the instrument industry. High-tempera-ture superconductors allow the development of magnetic field sensors, microwave devices, magnetic shielding instruments, and infrared sensors for defense applications. SQUIDs use Josephson junctions to build real-time instruments for measuring magnetic fields for possible applications in intelli-gence, surveying, research, and health. Instrumentation, as DeBresson sees it, is the ideal sector to improve the technical reliability of HTS devices. Next come the supercomputers, today's machine tools, where the search for hybrids combining superconductor and semiconductor characteristics may involve both LTS and HTS. HTS would then continue up the ladder of thresholds.

Even if DeBresson's prediction model presents an overly idealized and linear sequence ignoring likely unpredictabilities and surprises, it is a useful reminder of how many different kinds of hurdles or performance thresholds have to be overcome in a new technology's maturation process. But changes in the required performance criteria usually also involve a change in the composition of the teams working to attain them.

Thus, the adoption of an engineering approach in applied superconductivity research implied new performance criteria: potential users demanded a good return on capital investment. This was unattractive to the scientists who had worked on superconductivity in the early days. They either left the field or began designing equipment for clients with specific requirements. Typically, they worked in groups in government-funded or industrial labs, and only rarely in universities with strong external links. Only a third of the LTS basic research groups continued superconductivity research in the universities, usually in small enclaves outside the leading industrial countries. The Soviet Union (now the CIS) also continued its theoretical efforts; it had a long tradition in the field, and its scientific isolation was also an insulation from market constraints (Pippard, 1977; Waysand, 1987).

Of the numerous groups participating in the rapid rise of LTS research in the 1960s, many subsequently switched to other areas of scientific interest requiring theoretical knowledge and lab equipment similar to those used to study superconductivity. Smaller waves of scientific interest in phenomena related in some way to superconductivity followed, which connected parts of the research groups in a wider network, including groups working on organic superconductivity, spin-glasses, and heavy fermions.

Former superconductivity researchers thus still constituted a latent network in the "diaspora" that was activated with the discovery of HTS. They were joined by other groups formed in the 1950s and 1960s in response to the rise of a variety of basic research fields such as semiconductors and ferroelectrics. Others, notably Bernd Matthias and his students on both sides of the Atlantic, had maintained close links throughout the period and at the time of our study made up a large proportion of the HTS research groups in the United States and Europe.

Research fields thrive and shrivel against the backdrop of basic research groups forming and re-forming to address specific scientific questions. The researchers operate in a broader field of technological applications difficult to influence due to their systems nature. But the applications provide the rationale for setting up national research programs; basic research can then create new research niches or finds its niches being transformed or closed, by turning toward applications or toward other, more interesting niches elsewhere.

Today, this wider context for research is no longer one of pervasive and unquestioned technological optimism, nor are large technological systems considered invulnerable anymore. In the present situation of globalization, all scientific discoveries and technological innovations are faced with stiff international competition. The industrial connection is no longer optional, but indispensable – even for basic research.

2.3 The breakthrough: the discovery of high-temperature superconductivity

Superconductivity is a quantum-mechanical property in which electrical currents flow through a solid material with zero resistance. Since resistance is the conversion of electrical energy to heat (usually waste heat) and since the necessity to avoid heat damage is a constraint on applications, a superconducting electric motor can be up to ten times as powerful as a conventional motor of the same size, while consuming less energy. Conventional circuits require constant energy input, since resistance dissipates the entire current virtually instantaneously (on a human time scale). But closed superconducting circuits retain their charge in "perpetual motion"; this phenomenon is exploited in superconducting magnets and in energy accumulation circuits.

In spring 1986, when K. Alex Müller and Georg Bednorz demonstrated the occurrence of superconductivity in a new class of materials at "high" temperatures, "low-temperature" superconductivity had been known for some seventy-five years. Since the Dutch physicist Heike Kamerlingh Onnes discovered it in 1911, the phenomenon has occupied a special place in physics. Great minds have sought a theory explaining it and skilled experimentalists have tried to develop some of the many potential technological applications already envisioned by Onnes himself. But superconductivity proved extremely recalcitrant; the superconducting state is destroyed by the presence of relatively weak magnetic fields. Hopes for applications have repeatedly been raised, only to be dashed again. Although a number of labs around the world began research on superconductivity in the late 1920s and early 1930s, it long remained basic research. Theoretical work did not progress as hoped and practical obstacles to applications kept arising. Near the end of the 1930s, research in the field ceased almost completely, though for a variety of contingent reasons: Russian groups disappeared in Stalin's purges; the Dutch group lost the momentum imparted by its founder, Onnes; and the British were attracted to another phenomenon, superfluidity (Waysand, 1987; Ortoli & Klein, 1989).

After World War II, the development of a simple and relatively inexpensive device for producing liquid helium, the Collins helium liquifier, revived interest in low-temperature physics, including superconductivity. In the 1950s and 1960s, the number of publications on cryogenics grew by 10–15% a year. In *Physical Abstracts*, the number of papers on superconductivity shot up from a mere 66 in 1953 to a whopping 830 in 1966 (Pippard, 1977: 3). By 1960, 35 elements and a thousand different materials and alloys had been shown to exhibit some degree of superconductivity under specific conditions, compared with only 80 in 1935 (Ortoli & Klein, 1989). A year later, the first functioning superconducting magnet, with 60 000 gauss, was built.

Innovation in instrumentation renewed interest in superconductivity. The increased availability of liquid helium and the development of He^3 cryostats able to cool as low as 0.3 K were decisive. Low-temperature techniques had improved through the use of liquid hydrogen as a rocket propellant, the extensive industrial use of liquid gas separation to obtain oxygen, and developments in liquid gas transportation and storage. The new technology revitalized systematic measurements of the many strange magnetic, thermal, and optical properties of superconductors. Bernd Matthias, assisted by John Hulm and Ted Geballe (now at Westinghouse and Stanford, respectively), conducted a systematic "cartographic" study of the periodic table, allowing the formulation and testing of many empirical rules and leading to the discovery of materials with critical temperatures (the threshold under which the material becomes superconducting) as high as 23 K. These were "A_3B" intermetallic compounds of niobium with tin, silicon, or aluminum (where A = Nb and B = Sn, Si, or Al), or Laves phases, the most promising being lead, molybdenum, and sulfur. Their critical temperatures remained unsurpassed from their discovery in 1958 until 1986.

In the 1950s, it was found that there were two types of superconductors. Type I, mostly metal element superconductors, exhibited perfect electrical conductivity for direct current and perfect diamagnetism, i.e. they completely excluded magnetic flux from the material. Type II superconductors were not perfectly diamagnetic; they allowed magnetic flux, quantized in fluxons, to penetrate the material. Both concepts had been developed in the late 1930s in theories on thermodynamics and electromagnetism posited by Ginzburg and Landau as well as by the London brothers. Type I superconductivity vanishes in the presence of weak magnetic fields, rendering it unpromising for applications. Type II superconductors were more interesting, since they withstand magnetic fields of more than 1000 oersted, i.e. 1000 times the strength of the Earth's magnetic field. But another requirement is hardness in response to current-carrying in a magnetic field; that is, the fluxons must be pinned to the material rather than moving in it as a current. By the early 1960s, this trait was found in several intermetallic compounds and in some niobium–zirconium and niobium–titanium alloys.

Meanwhile, theoretical interest in superconductivity had been revitalized by new experimental findings, particularly that the isotope composition of a metal influences its critical temperature. The most important theoretical advance came in the United States with the publication of the Bardeen, Cooper, and Schrieffer (BCS) theory in 1957, which explains why electrons pair up in superconductivity and why they react to magnetic fields and temperature as they do. It crowned a series of pioneering theories proposed by Ginzburg, Landau, Abrikosov, Fröhlich, and others and provided a comprehensive view of a number of macroscopic thermodynamic phenomena. The BCS theory attracted considerable media interest: the *New York Times* heralded it with the headline "The Discovery of a Superconducting Alloy at Room Temperature is now Imminent" (Ortoli & Klein, 1989: 1957). The BCS theory helped scientists understand why there are two types of superconductors – but did not predict what material would fall into which category, nor did it provide a clue for finding better superconductors. It was not without predictive power, though, having led to the discovery of coherence phenomena such as the Josephson effect.

In 1962, Brian Josephson, then a young graduate student at Cambridge (UK), solved another crucial puzzle about superconductivity. By combining tunneling theory and the general concept of weak links, he created the theoretical basis for a device now known as the Josephson junction, a sandwich of two superconductors separated by a thin insulating barrier. The Josephson junction opened up a new branch of applications for superconducting technology, particularly in microelectronics, but also in devices such as the superconducting quantum interference device (SQUID), a highly sensitive detector of magnetism.

The era of high-field superconductivity had begun, a lab phenomenon finally having turned into a viable technology. Many of Kamerlingh Onnes' original dreams of industrial applications now seemed closer to realization. Superconducting electromagnets, for example, would provide high fields while consuming virtually no electric power – and without requiring radically new engineering; and power plants and transmission cables cooled to liquid helium temperatures would operate without losing current due to resistance. But in reality, much work remained before superconductivity could move out of the laboratory; the first superconducting magnets suffered major breakdowns and there were other failures. Further technical advances came with cryostatic stabilization. The first commercial superconducting wire was produced in the early 1970s. Superconducting magnets found their first uses in other labs, for example as components in particle accelerators. The most successful utilization of low-temperature superconductivity came unexpectedly in the early 1980s. Superconducting magnets enabled the development of nuclear magnetic resonance imaging (MRI) devices, a great step in medical diagnostics.

In 1973, superconductivity was found in oxides at 13 K, demonstrating that the field still held surprises. But for the time being, researchers were unable to raise the critical temperature in superconducting oxides, and research turned elsewhere. Thus, until the Zurich discovery, 23 K was the highest critical temperature of any known superconductor. There was a seemingly satisfactory theory explaining the phenomenon, but experimentation was yielding few advances. As a field of basic research, superconductivity seemed closed and interest had slowly waned. It seemed to have entered the phase of applied technology for good, and the two main areas of application date from this period: the use of zero-resistance in power transmission, generators, and high-field magnets; and the use of coherence phenomena in microelectronics. But these successful applications fell very short of earlier claims that superconductivity would be used to transmit electricity to households. More manageable materials were available for many commercial applications; other potential applications seemed to depend on progress in other fields, such as fusion. Commercial competition from existing, matured technologies was fierce. As a result, several major research projects were scaled down or canceled in the 1970s and 1980s.

Müller and Bednorz's unlikely discovery

That was the situation in 1983, when, at the IBM Rüschlikon Laboratory in Zurich, Switzerland, Alex Müller and Georg Bednorz began investigating ceramic oxides, seeking new superconducting materials with higher critical temperatures.

In 1982, after working for fifteen years as a senior research manager at Rüschlikon, Müller was appointed an IBM senior research fellow. The appointment can be seen as a pre-retirement reward from his employer, giving him more time to pursue his own work. For most of his professional career, Müller had worked on strontium titanate ($SrTiO_3$) and related perovskite compounds. He had gained an international reputation for his work on ferroelectrics and, later, on critical phenomena in structural phase transitions, i.e. changes in the spatial arrangement of atoms in solids that determine their physical properties. But not until a twenty-month sabbatical spent at the IBM Yorktown Heights Laboratory in 1980 did Müller become familiar with the intricacies of superconductivity research. He was particularly fascinated by the IBM Josephson computer project, a $ 300 million effort to build a fast computer based on superconducting switches. The computer project was later abandoned, but the knowledge that several complex oxides were superconducting led Müller to believe it might be possible to find new compounds with unprecedented critical temperatures.

Returning from the US in 1983, he asked Georg Bednorz, a young crystallographer also working at the IBM lab in Zurich, to collaborate with him in searching for superconductivity in oxides. Bednorz had already been briefly exposed to work on the superconducting $SrTiO_3$ and agreed at once. In late summer 1983, they began seeking high-temperature superconductivity in the LaNiO system, since it fitted into what they called their guiding concept. They had been influenced by studies of the formation of polarons through the Jahn-Teller effect, which they thought would lead to stronger electron–phonon coupling and hence to higher critical temperatures. In 1985, after a series of disappointing trials, Müller and Bednorz considered abandoning their search. Bednorz says they probably continued only because their experimental situation improved. They switched to a new series of compounds, but still observed no signs of superconductivity. The turning point came in late 1985, when Müller ran across an article by the French scientists C. Michel, L. Er-Rakho, and B. Raveau, who had found a BaLaCu oxide with perovskite structure that exhibited metallic conductivity at a relatively high temperature (Michel *et al.*, 1985). This oxide fitted Müller's and Bednorz's concept. They tested a sample, soon witnessing a sudden drop in resistance. They were able to push the threshold up to 35 K, much higher than the highest earlier critical temperature previously observed. This was the breakthrough they had sought! (Bednorz & Müller, 1987; Müller & Bednorz, 1987)

They decided to publish their findings even before performing the crucial magnetic measurements that could conclusively confirm the presence of superconductivity by demonstrating the Meissner–Ochsenfeld effect, the expulsion of an applied magnetic field. Bednorz approached his manager, a member of the editorial board of the journal *Zeitschrift für Physik*, and the paper was submitted for publication in April 1986. It appeared in September, cautiously titled "Possible high-T_c superconductivity in the Ba–La–Cu–O system" (Bednorz & Müller, 1986). The SQUID magnetometer they had ordered had meanwhile arrived. Together with a visiting Japanese colleague, they performed the measurements and demonstrated the presence of the Meissner–Ochsenfeld effect. (Bednorz *et al.*, 1987a) The new era of high-temperature superconductivity had begun.

Müller's and Bednorz's discovery of HTS violated expectations and conventional scientific wisdom in at least three ways: the discoverers themselves; the site of the discovery; and the scientific ideas involved.

The discoverers were relative outsiders to the field of superconductivity. Müller was an experienced, acknowledged specialist on perovskites; Bednorz was a crystallographer. As Bednorz later remarked, for someone without a background in the physics of oxides and not directly involved in pushing critical temperatures to the limit, casual observation of advances so far would have led him to believe that intermetallic compounds held no further promise, since the current record T_c, 23.3 K, had stood since 1973 (Bednorz & Müller, 1987).

Of IBM's three superconductivity laboratories, Rüschlikon was by far the smallest and most modestly equipped. The odds seemed stacked against it, compared with other large and booming labs like Bell, where the concentration of first-rate researchers included Cava and Batlogg, seasoned superconductivity specialists. At large-scale industrial research labs like IBM Yorktown Heights and Bell, ideas are mass-produced in an atmosphere of intense stimulation and intense competition that pushes researchers to their limits. Such labs have been described as places where hunters are set loose to catch the ideas formulated by more patient gatherers.

By contrast, Rüschlikon resembles an idyllic retreat. Small in scale, its research teams of experienced older managers and brilliant young researchers pursue their ideas in relative freedom. Gerd Binning, who shared the 1986 Nobel Prize with Heinrich Rohrer for their work on the scanning tunneling microscope, has described Rüschlikon as his ideal research management structure. The laboratory is divided into three groups, each in turn divided into three subgroups of about three researchers with an optimal flow of information and communication (Binning, 1989: 237–8). Like Müller and Bednorz, he finds the close, personal interaction of small, collaborative teams conducive to the free pursuit of ideas that is the basis of scientific creativity.

The third way in which the discovery of HTS was unexpected was that it contradicted received opinion in the field. It not only overturned the established empirical rules of Bernd Matthias, one of the grand old experimentalists in superconductivity, it also unveiled previously unknown phenomena not accounted for by the BCS theory. The BCS theory had seemed to offer a reasonably rigorous and satisfactory explanation of the known behavior of superconductors; its recognition through the award of the 1972 Nobel Prize for physics had made superconductivity seem a consolidated and relatively quiet area of solid state physics (Gavroglu & Goudaroulis, 1989). This was confirmed in 1986 by a flush of retrospective articles on the occasion of the seventy-fifth anniversary of the discovery of the phenomena. Theoretical and experimental work had resulted in a belt of solid technological applications and the growth of a considerable worldwide market for superconducting magnets and

Josephson junctions. The field seemed to have left an "adolescence" of exaggerated and unconfirmed claims behind it, and to be resistant to further surprises. This is why Müller and Bednorz were initially very cautious in interpreting their results. This caution – and the coincidence that their new compound became superconducting at 30 K, a spectacular temperature, but one not yet ruled out by the BCS theory – may have contributed to the credibility they enjoyed after a brief initial period of skepticism.

2.4 The euphoric phase

It took some time before the findings of Müller and Bednorz achieved their full impact. Perhaps this was because they published their paper in the *Zeitschrift für Physik*, a reputable German journal, but not a forum for important new findings like *Physical Review Letters* or *Europhysics Letters*, which also have shorter publishing times. A highly respected superconductivity researcher and consultant to the German Ministry of Research and Technology told us in an interview that on receiving the Müller and Bednorz paper, his first impulse had been to throw it away. Many established scientists in the field reacted similarly; over the years, they had grown used to spurious announcements that higher critical temperatures had been achieved. Sooner or later, all had proven unfounded. There had even been cases of reports being published though their authors had done no empirical work. The highest confirmed critical temperature had remained constant for more than a decade, and it was no longer considered worth the effort to duplicate experiments claiming new records. Even Müller and Bednorz themselves "expected that confirmation and acceptance ... could take as much as 2 to 3 years" (Müller & Bednorz, 1987: 1136).

Thus prior to confirmation of the Meissner–Ochsenfeld effect, there seemed little reason to make an exception for Müller and Bednorz. Müller was a specialist in perovskites, not superconductivity.

But other international groups working on perovskites noted the Rüschlikon results. They knew and respected Müller as a persistent and meticulous researcher. The groups around Shoji Tanaka at the University of Tokyo, Zhao at the Physics Institute of the Chinese Academy of Sciences, Paul Chu at the University of Houston, and Robert Cava and Bertram Batlogg at American Telephone and Telegraph Company (AT&T) Bell Labs. were all familiar with ceramic oxides. Tanaka's group was the first to reproduce the Müller and Bednorz results, a confirmation that "fanned the fire in the United States" (Müller & Bednorz, 1987: 1137). This was soon followed by detailed structural analysis of the material at Tokyo, IBM, and Bell Labs. The new high-temperature superconductor apparently contained three phases, only one of which was superconducting. It showed a structure familiar to any crystallographer – a tetragonal, layered perovskite (K_2NiF_4-type) – and fueled speculation that this atomic structure was the key to superconductivity.

What we know about the work of Chinese groups is limited to a few mentions in US newspaper articles and scientific papers. By contrast, the story of Paul Chu's group at Houston was soon plastered across the headlines. A science journalist also gave a heroic description:

Reaching his office, Chu switched on the light and glanced at his functional metal desk to assess the damage. Without so much as a cup of coffee to spur him on, he settled himself in his swivel chair and began to work. As it turned out, he didn't get beyond the first few items that morning because there, near the top of the stack where his graduate student Li Gao had placed it the previous day, was a freshly copied, five-page article from *Zeitschrift für Physik*. Chu could barely contain his excitement as he re-read the title: 'Possible high-T_c superconductivity in the Ba–La–Cu–O system'

(Hazen: 1988 24).

Chu did attempt to reproduce the IBM results, reaching 30 K for his BaLaCuO samples. Inspired by his experience with high-pressure effects, he discovered that the critical temperature could be pushed to 40 K, a result he submitted to *Physical Review Letters* on December 15, 1986. He was convinced that other materials must exist with even higher critical temperatures (Chu *et al.*, 1987).

Even before Chu's high-pressure experiments, a systematic search for new materials with similar structure had already begun in late 1986. Working independently, all the groups involved tried replacing barium with other elements of the same class, all choosing strontium to create an isostructural phase in a different system. In late December 1986, Tanaka's group and the team of Cava and Batlogg submitted reports of superconductivity in LaSrCuO at 35 K and 36 K, respectively (Cava *et al.*, 1987). A few days after reading a December 31, 1986 *New York Times* front page article on the Bell Labs. result, Chu amended his high-pressure paper to report that his team, in collaboration with Mau-Kuen Wu's group at Alabama, had also reached 42 K in the LaSrCuO system. To remain in the forefront, he even claimed to have observed signs of superconductivity, however unreproducible, at temperatures higher than 56 K. Chu's result was rapidly followed by a group at Bell Communication Research Laboratory (Bellcore), who, on January 21, 1987, submitted a report to *Science* on superconductivity at 40 K in the strontium compound, and also by Bednorz, Müller and Takashige, who announced their results in February. Finally, the Beijing group also reported success in this direction (Tarascon *et al.*, 1987; Bednorz *et al.*, 1987b).

Thus, by the last months of 1986 and in January 1987, the situation had changed completely. Not only had Müller's and Bednorz's results been reproduced and widely accepted, the detection of even higher critical temperatures in other materials suddenly bestowed unprecedented prestige on the new, very "hot" line of research. In the "bestseller" list of most frequently cited scientific papers, published regularly by the Institute for Scientific Information, the Müller and Bednorz (1987) paper ranked first, with 4168 citations in 1987, followed by the M. K. Wu *et al.* (1987) paper, cited 3331 times, and H. Maeda *et al.* with 1783. Insiders clearly saw that something unique was happening, and everyone wanted to be among the leaders in the race.

Further developments took place in a climate of fierce competition. The excitement of the IBM discovery spread rapidly through scientific circles as more and more groups joined the feverish hunt for new materials. This time the Houston–Alabama collaboration was first, due to its ingenious idea of replacing lanthanum with the rare earth element yttrium and to the Alabama group's expertise in preparing ceramic oxides. In early January 1987, Paul Chu found YBaCuO superconducting at about 90 K and immediately applied for a patent on the new compound. But, not submitting his paper to *Physical Review Letters* until February 6, Chu was confronted with a major problem inherent in the conventional scientific publishing system: the possibility of information leaks or publication delays. He chose unconventional tricks to protect his work and ensure its priority. First, the paper contained two systematic mistakes that rendered it useless to any reader. He waited until February 18, the last possible moment before publication, to send his corrections to the journal (Wu *et al.*, 1987). Second, fearing another team might meanwhile find "his" new compound on its own, he held a press conference on February 15 to announce, without providing details, that he had discovered a new material with a critical temperature near 98 K.

The scientific community and the lay press responded ecstatically to Chu's publication. Müller and Bednorz had made the conceptual breakthrough in superconductivity, but the step from 23 K to 30 K in a new class of materials promised no practical advantage, since liquid helium – expensive and difficult to handle – would still be the required coolant. Chu's detection of superconductivity in YBaCuO at 98 K, on the other hand, was a technological breakthrough; the critical temperature could now be attained with liquid nitrogen, which, as the *New York Times* February 16, 1987 noted, "cost a tenth as much, was twenty times as efficient, and was much loss volatile." Speculation about commercial applications immediately flared up and ranged from the generation and transmission of electricity through medical diagnosis to powering trains with magnets (Sullivan, 1987). But the question of coolants was not the only stimulating aspect of Chu's finding. He had discovered another new class of superconductors, different in structure from those previously known, again hinting at unimagined possibilities.

Amid this growing excitement, the American Physical Society (APS) scheduled a "Special Panel Discussion on Novel Materials and High-Temperature Superconductivity" during its annual meeting in New York on March 18, 1987. Only a few months had passed since the publication by Müller and Bednorz and only a handful of physicists had checked the results and really understood their importance, but almost 3000 physicists squeezed into the New York Hilton to hear the latest news. An unforgettable event unique in the history of physics, the conference later came to be known as the "Woodstock of physics" (Robinson, 1987).

The talks were brief, and the first round of speakers included members of the five leading groups in the field of superconductivity from IBM, Tokyo, Houston, Beijing, and Bell Labs. They were followed by contributions from theoreticians and experimentalists from other labs. It turned out that nearly every experiment and idea had been duplicated during those hectic months of 1986 and 1987. But several puzzles remained, such as the structure of the new compound Chu had detected. The BCS theory, which had seemed to account for all phenomena linked to superconductivity, was no longer adequate. Either a new theory was needed, or the existing one had to be modified to cover HTS. Participants discussed the potential applications of the new materials and arrived at a consensus that the scientific importance of HTS would be matched in technological impact only if the materials could be made to carry strong enough electric currents and if usable forms of the superconductors were not too difficult to manufacture.

The APS meeting not only opened up unexpected scientific perspectives, it also moved superconductivity from the backwaters to the limelight. The entire field was shaken by abrupt and radical changes, especially in the modes of personal communication within and beyond the scientific community. The conference affected funding and mobilized public opinion.

Indeed, the discoveries by Müller and Bednorz and then by Chu initiated a chain reaction of interest. The number of scientists involved in superconductivity research increased almost daily. They came from a variety of disciplines – chemistry, solid state physics, and crystallography – and equipped with varying specialized skills and experience. Not all who joined the venture in its initial flurry were destined to do good research and remain in the field. Some were motivated by the prospect of increased research funding, others attracted by the field's sudden scientific prestige. Some were curious to duplicate results, while others dreamt of also finding a new material with an even higher critical temperature. For awhile, "creative cooking" superseded serious chemical synthesis.

Patterns of communication changed rapidly. Many hastily-written articles of dubious quality were published; new journals were launched to speed up the publication process, sometimes abandoning any formal refereeing system; expensive commercial newsletters were marketed to wide circles of subscribers; and scientific results were frequently announced to the general press before being published in scientific journals. Numerous conferences were held around the world.

To protect claims to priority over rival research teams and even close colleagues, some researchers developed a mania for secrecy. This was not only a matter of scientific recognition, but also of securing patents for future applications. Competition developed not only between research groups, but also between nations, especially the United States and Japan. Everyone wanted to get in on the ground floor of any emerging applications. A kind of gold fever had broken out in the research community; after years of disillusionment, unfounded reports, and stagnation, suddenly it was again possible to "do physics" and to dream of superconductivity at room temperature.

Through its intense media coverage, superconductivity had also captured the public interest. It was described as the scientific finding whose technological applications would sooner or later transform everyone's daily lives. The media promised loss-free power transmission and magnetically levitated trains as the forerunners of a new technological golden age. Superconductivity was played up as "little science" accessible to all: even school children could concoct superconducting compounds, and liquid nitrogen was available for levitating a sample above a magnet. Further research seemed destined to put applications within everyone's reach.

The excitement peaked in spring 1987. Many of the scientists we interviewed spoke of encounters with non-scientists who wanted to learn more about HTS, or of unprecedented publicity through interviews in the local press, TV, or radio. Several levels of interest reinforced each other.

The first level was scientific. As we have seen, a relatively settled area was "heating up". Theorists and experimentalists could challenge each other again. Interdisciplinary research had lacked subject matter in the past, but could now be tested in practice. New and rewarding forms of collaboration, particularly between physicists and chemists, loomed on the horizon.

The second level was technological. Media speculation about potential applications of HTS revived old euphoric dreams of a technological utopia. More cautious feasibility studies proliferated, frequently containing timetables predicting which concepts would be commercially realized, and when. Apart from power transmission, microelectronics was considered the most interesting and feasible field for applications in the mid- to long-term. New hybrids of semiconductors and superconductors were predicted. The transistor was invoked as a warning to expect the unexpected.

Finally, the euphoria spread even to leaders of national politics, particularly in the United States, where political and economic rhetoric were pushing for support for industry in its economic and technological competition with Japan. Any major advance in materials science could become a key area of technology and improve market performance. And any technology that contributed to solving energy problems was attractive, quite apart from considerations of competition. At public meetings, politicians underscored the links between scientific advances and technological innovation. On July 28, 1987, a US federal conference on the commercial applications of superconductivity was held, but excluded all European and Japanese researchers. President Reagan exhorted the nation to seize the opportunity created by HTS research. He announced an eleven-point Superconductivity Initiative: "Science tells us that the breakthrough in superconductivity brings us to the threshold of a new age ... It is our task ... to herald in that new age with a rush ... It's our business to discover ways to turn our dreams into history as quickly as possible." (Heppenheimer, 1987)

Things began to calm down, at least in scientific circles, in October 1987 when the Nobel Committee awarded Müller and Bednorz the Nobel Prize for physics for their work in early 1986. The award recognized the significance of their discovery of the first ceramic material to superconduct at the then-high temperature of 30 K. At the same time, the granting of the Nobel Prize meant one stimulating race was over. The anomalous situation of 1987 began to yield to calmer and more serious research. Many scientists who had merely jumped on the bandwagon returned to their previous work. It had become clear that years of serious research would be needed to understand the phenomena and that more years would pass before applications became commercially viable. The media, however, continued to cover the subject for some time. Politically as well as scientifically, the ground had been prepared for the establishment of national research programs, though much organizational effort would still be needed to launch them.

2.5 The return to normality

In early 1988, discoveries of superconducting materials with ever higher critical temperatures began tapering off. At an international conference at Interlaken, Switzerland, general enthusiasm was still palpable, but so were the first signs of returning caution. Ovchinnikov's report in *Physical Review Letters* claiming to have reached 240 K could not be confirmed, but had a kind of dumping effect on the research community. New avenues were still being explored, such as substituting fluorine for oxygen, but with little success. As a substitute to improve the overall properties of a superconducting compound, thallium had yielded the maximum critical temperature of 125 K (Sheng & Hermann, 1988). This five-degree improvement over bismuth was too weak an incentive for researchers to work with the extremely poisonous substance. By the end of 1988, it was accepted that the desired results would be found among the copper oxides and that critical temperatures beyond 125 K were probably not just around the corner. The giant steps gave way to modest progress.

Parallel to the experimental discoveries, theorists also proposed competing explanations for the phenomenon, either elaborating the BCS theory or offering radically new ideas. Müller's intuition proved resistant to quantitative theoretical description. The chemical complexity and poor reproducibility of the materials also made it difficult to subject hypotheses to rigorous experimental verification. Critical temperatures were more easily reproduced in the new ceramics than in the old superconductors, but other data, such as those on specific heat, forbidden gap (optical or electrical), or current-carrying capacity could not always be reproduced – even within the same laboratory. There is still no satisfying theory that covers all forms of superconductivity (Anderson & Schrieffer, 1991).

The development of semiconductors can help to put the efforts toward a theory and the practical utilization of superconductivity into perspective. The prerequisite for the production of pure single crystals (a transistor is a pure single crystal with controlled doping) was the development of Pfann's zone-melting process, which allowed intrinsic properties to be distinguished from those dependent on impurities. Fortunately, the chemical composition of useful semiconducting materials is relatively simple. Production procedures for semiconductors have meanwhile reached a high degree of perfection, but in superconducting ceramic compounds, the equilibrium between the elements and the volatility of oxygen are still poorly understood and largely uncontrollable. The first steps toward precision in producing them were made in refining physical thin-film deposition techniques, rather than in producing bulk materials. It proved easier to control thermodynamic conditions during the off-equilibrium process of condensing small amounts of evaporated elements onto cold substrates than in the melting or solid state reaction of larger quantities under equilibrium.

Despite these practical difficulties, the two years following Müller's and Bednorz's discovery brought a number of conclusions about superconductivity. First, the majority of researchers now believed that superconductivity carriers are pairs of electrons, as described in the BCS theory. Second, the primary difference between high-temperature and conventional superconductors is that the coherence length of the former is of the order of magnitude of the crystal lattice, while in low-temperature superconductors it is 200 to 1000 times greater. Third, superconductivity is closely related to the copper oxide crystal planes and has anisotropic properties. These characteristics threw some light on the theoretical and experimental problems.

On the theoretical side, the techniques used to understand the local properties of condensed matter are not yet as well developed as those that can use averages of large numbers of atoms and electrons: concepts such as Fermi surfaces or electron distribution lose part of their significance for HTS. Some experimental data indicates that it is unlikely that critical temperatures can be raised much further if this depends on another reduction of coherence length.

On the experimental side, some radical differences have been found between the behavior of conventional and high-temperature superconductors. First, critical temperature is no longer believed to depend on the average behavior of many crystal lattice cells, but is apparently determined by what happens in good superconducting cells. For a conventional metal alloy superconductor, variations in the critical temperature may be the result of impurities or deformations. In HTS, even a dirty compound shows the critical temperature of its good molecules, if there are enough of them. Thus, a high critical temperature alone was no longer a good indicator of the quality of a superconductor.

A second difference lies in current-carrying properties. The interconnection, i.e. the quantum coherence between one crystal cell and the next, has proved fragile in HTS materials, easily weakened by impurities or anomalies. This in turn indicates that the observed critical current depends strongly on the quality of the material. Only well-grown epitaxial films or single-crystal samples show reasonably high critical currents, and an applied magnetic field can easily uncouple adjacent grains even in the best materials. Yet good interconnections between grains are not sufficient to ensure current-carrying capacity; experimental attention has recently focused on the new materials' hardness to magnetic fields, produced by the "pinning" of magnetic flux lines.

These findings and the practical difficulties of material preparation partly explain the return to "normal science" and systematic research – the incremental accumulation of evidence or test results and their comparison with hypotheses and calculations, an iterative learning process allowing the gradual selection of scientific and technological routes to a fundamental understanding of the phenomenon.

It would go far beyond the scope of this book to describe all the materials that were discovered showing superconductivity at high temperatures, but two examples will serve to illustrate the fundamental change that had taken place. Most of the superconductors found in the first wave of investigations were based on copper oxide planes – and it was thought for awhile that this was the clue to a theoretical explanation – but a systematic search was also begun in other new directions. Indeed, the HTS surprise led scientists to test every new conducting material for superconductivity, which would have been inconceivable only two years earlier. Thus, for example, investigations of non-copper oxide superconductors turned up a bismuth-based compound (BaKBiO) with a critical temperature of 30 K (Cava *et al.*, 1988). And in 1991, a completely new kind of superconductor was discovered: K_3C_{60}, a so-called buckminster-fullerene, a "molecular" high-temperature superconductor at up to 30 K with a coherence length near the molecular size.

At the time Müller and Bednorz made their discovery, the common feeling of the researchers was that past theoretical knowledge and the practical applications of superconductivity would rapidly help find an explanation of the microscopic mechanisms and lead to the practical exploitation of the new discovery. Now, nine years later, and with many thousands of person-years of research accumulated, the situation resembles that experienced with the low-temperature variety of superconductivity, which for decades resisted the equally highly qualified efforts of top scientists to understand it.

But there are substantial differences in the scientific and practical situations, not to mention the economic and social environment, which is far more complex and sophisticated than the pre-World War I Europe of Kamerlingh Onnes. Starting with applications, there is a greater variety of niches for HTS than there once were for LTS, and several commercial demonstration devices already exist. For example, HTS has been successfully applied in the electrical energy field in the use of reactive elements to ensure the quality and continuity of power in electrical lines. Another tested application is HTS elements in the expansion of cellular communication and data systems, a potential high-growth market where high-temperature superconductors have properties giving them a competitive edge over semiconductors. Smaller specialized niche applications have been realized in the market for LTS components, including current leads for MRI magnets, sensor elements for diagnostics in helium-cooled systems, and SQUID detectors operating above liquid helium temperature.

But HTS applications have not expanded as expected, because the development of large energy-related uses has been delayed by the very slow development of usable cables and wires with current-carrying capacities unaffected by magnetic fields and mechanical treatments like winding and stretching. This slowness is due to the very complex and inhomogeneous nature of the superconducting compounds found so far. So research will have to explore many different avenues and make small, incremental advances, each difficult to incorporate in a practical industrial process.

One specific hope of the early phase has not materialized: no unforeseen application has emerged to "leapfrog" the difficulties and create a market large enough to fuel research, as MRI did for LTS.

Nearly a decade after the breakthrough, the bandwagon effect so evident in the first wave of research activities persists. Members of the HTS community judge many of the papers still produced on the subject to be repetitive, reproducing known results with minor or insignificant variations and raising no important new questions. Several thousand papers deal with the hope of finding higher critical temperatures by doping and treating known compounds in various physico-chemical ways. There has been a steady flow of articles detailing the pinning and release of flux lines, often in insufficiently characterized materials. Could this avalanche of papers be related to the conviction held by funding agencies and research groups that HTS should be closely observed, in case of a sudden new development or further significant breakthrough? One of our interviewees remarked that this overproduction of papers is even promoted by scientific journals' willing acceptance of articles that are far from being innovative.

In this wealth of papers, few researchers attempt to measure significant parameters of microscopic mechanisms in well-characterized samples, and the relation between theory and experiment is unsatisfactory. Most experiments never deal with the full implications of a theory, but try to prove or disprove minor predictions. Theories, for their part, are evolving incrementally from the two or three most insightful suggestions that appeared in the first years. Theories and experiments attempting to unravel the basic mechanisms of HTS seem to run in circles; perhaps all the ideas have been exhausted for the time being, and can only be reshuffled to produce minor variations in slightly different contexts.

It seems clear that the theoretical techniques and probably some of the concepts developed in solid state and condensed matter physics to reduce the complexity of an assembly of hundreds of billions of interacting atoms to solvable equations are not powerful or accurate enough to illuminate the origin of high-temperature superconductivity in a quantitative and predictive way. It is now known that, within an ensemble of atoms no larger than a single cell, superconductivity arises out of a seesaw balance between insulating and conducting states, and between antiferromagnetic order and magnetic disorder. In the coupling of electrons, high-temperature superconductors resemble their LTS counterparts, but the electron pairs seem to arise due to a self-organization of magnetically interacting atoms and electrons, rather than from the elastic deformation of the crystal surrounding the electrons. Thus, some characteristics point toward a magnetic interaction – which, in LTS, had been found to be antagonistic to the emergence of the superconducting state! Bernd Matthias, one of the leading figures in LTS research, was fascinated for years by this new coexistence of magnetism and superconductivity.

Perhaps a successful theory would combine approaches that currently seem opposed. More study of HTS materials *above* their critical temperature might prove revealing, even though the complexity of the materials poses a tremendous challenge.

3

Reconfiguring actors and knowledge: the organization of a new research field

The emergence of HTS as a research field is an example of how positing a situation can make it real. Discourse and beliefs, rhetoric and persuasion, and a vision of a bright technological future – hardly supported by reliable facts at the time – acted as a catalyst. As the concept of "windows of opportunities" suggests, when new technologies appear on the market, new opportunities suddenly seem to exist, but the period in which they can be realized and exploited is brief (Perez, 1983; 1989). In the end, unsurprisingly, there are winners and losers. But while institutions maintained their structural grip and while path-dependence and varying degrees of preparedness had their effects, for a short, compressed time, scientists' vision and rhetoric, policy constructs and persuasion succeeded in collusively reshuffling some of the more inert parts of the science system, before they resettled into the familiar pattern of institutional stability.

The emergence of a new research field underscores that the science system is not set once and for all; knowledge of its history is thus an essential prerequisite for understanding it: "The passage of time, and changes it brings in the factors and phenomena that interest us, are our single best resource" (MacKenzie, 1990: 7). The study of a process of change is hardly in danger of mistaking a moment for an eternal condition. But it is difficult to distinguish a unique event from more enduring developments that permit generalization. The participants we interviewed, the institutions we visited, the situations and choices reported to us, and above all the state of scientific and technological knowledge continue to change. For those who want to follow it, the development of HTS is an especially rapidly moving target. There are as yet no canonical versions of what has happened or in what sequence, nor of the accounts offered by the main actors or of the contributions of the many nameless ones. The more recent discovery of the buckminster-fullerene superconductors and the fabrication of first small HTS devices show that, even if the expected boom in technological innovation is not yet in sight, the field is nonetheless far from closed.

Though not completely settled, it has certainly calmed. After the wild utopian hopes engendered by the spectacular take-off, researchers' anomalous behavior reverted to more routine patterns. Once funding agencies and industry had made their decisions, assessments of the technological potential of HTS grew more sober. The media moved on to other "hot" topics. Some institutional arrangements were altered; new research consortia and superconductivity centers were set up, achievements seemed variously modest or disappointingly minor. Assessments of the international context and the effects of the discovery on the economic competitiveness of the major industrialized countries in high-tech markets have also vacillated. We may have witnessed the establishment of what Rustum Roy called "a worldwide empirical science production machine" ensuring that "there will continue to be a finite number of large and small discoveries always made serendipitously, i.e. by careful, well-trained observers capitalizing on the ubiquitous chance events"(Roy, 1988: 33–4).

But even if its take-off is unexpected, the trajectory of a moving target begins in preconditions. Before HTS, low-temperature superconductivity had also once inspired high hopes for technological advances; these hopes were still remembered by a whole generation of researchers. The failure of nuclear energy to develop as forecast had clearly underscored the vulnerability of new system components – like conventional superconductor applications. While researchers waxed enthusiastic, national policy-makers displayed greater caution in the amounts of research funds allocated, and industries, measured against their total research potential, were certainly the most cautious of all.

The phenomena of superconductivity was first noted about three-quarters of a century ago. Many interesting theories have been constructed to explain it, and it has been variously conceived and measured (Gavroglu & Goudaroulis, 1989; Ortoli & Klein, 1989; Dahl, 1992; Ott, 1992). Its history is part of the labyrinth of pathways of inquiry that have opened up, sometimes arriving at a goal, sometimes proving to be blind alleys, and sometimes left incompletely explored. A range of problems regarding superconductivity has been declared solved. Some stretches of this historical labyrinth are no longer of interest except to historians of science or of technology.

Müller and Bednorz opened up new segments of the labyrinth with their discovery of HTS. Individual decisions, motivations, and circumstances that led researchers to move into the new field mask their collective relation to the ups and downs of other research fields. The intellectual biographies of individual researchers and the fate of their networks reveal manpower to be a frequently

overlooked essential factor in the dynamics of research fields. The sudden acti-
vation of latent research networks and the scientific knowledge embedded in
them, a bandwagon effect, and the attraction of problems prematurely declared
solved all pulled a large variety of researchers into the field. There were veteran
experts from low-temperature superconductivity; theorists, for whom the
switch was probably easiest; and newcomers from other disciplines, who
brought varying research skills and experience. The variety of their institutional backgrounds gave them different starting positions. This heterogeneous
mixture of disciplines and expertise is still reconfiguring to become the knowledge base of the new field.

Reconfiguration of knowledge and people is not automatic. It is contingent
on starting positions, beliefs and perspectives for the future. All research programs, Krohn and Küppers remind us, are perspectivistic constructions of future knowledge. While publications relate to past achievements, research proposals state a specific intended output for the future (Krohn & Küppers, 1989:
87–95). As the basis of its negotiations for funds, a research group setting up a
research program underscores the promise of an avenue of investigation; national research programs are the ensemble of research proposals submitted by
research groups and accepted as containing sufficiently credible perspectivistic
constructions of future knowledge and outcomes. This officially validated
package of probable future knowledge is the bargain struck between researchers on the one side and policy-makers and funding agencies on the other.
All national research programs in HTS were premised upon more or less detailed claims about expected achievements. Policy-makers agreed to believe the
promises offered them, even though any claim to future knowledge necessarily
begs the question.

From a sociological perspective, the establishment of HTS as a new research
field was a complex process. The many overlapping and intersecting layers become increasingly inseparable as the story unfolds. The extended lab of HTS
research begins at the level of individuals acting in accordance with their personal interests, driven partly by scientific curiosity, but also by such down-to-
earth factors as new financial resources, career wishes, the hope of finding a
niche to work in, etc. To reinforce their positions, the researchers form strategic
alliances with other individuals or groups in different institutions offering stability and structural support. But these institutional settings follow a logic of
their own, imposing certain choices, setting the structural preconditions, and
making specific kinds of alliances necessary. All this takes place in a policy
space where discussion on how to increase the technological potential of scientific research is heating up. A further layer of complexity is added by the fact
that these researchers also moved in a public space created by the euphoric

press coverage of HTS. The public had become an important, if hard to perceive, actor in this newly opened hybrid space of policy debate.

Which main actors joined together on the scientific level? What were their institutional backgrounds, and to what degree was the constant rhetoric of interdisciplinarity a reflection of the reality of everyday research? What led scientists step into a new field, promising but largely unknown? What efforts were made to structure the research environment to allow the creation of new research programs and to channel the flow of money into the field? Which researchers could claim expertise in a field lacking clear standards? As a further central issue, we shall examine the major changes in social interactions – i.e., communication, cooperation, and competition – within the newly created scientific community. Finally, some facets of everyday life we observed in our short visits to the three labs color the picture; though these are too small a sample to suggest general validity, they do add valuable information on how research concepts meet research realities.

3.1 Reconfiguring in different institutional settings

The initial overwhelming reaction to the discovery obscured the motives of the scientists, who came from different disciplinary backgrounds. They had a variety of motivations, starting positions, and infrastructural possibilities, and were embedded in varying national and policy contexts. Their research goals could range from answering fundamental questions about the functioning of superconductivity to the construction of artifacts incorporating this new knowledge. Above all, they belonged to different institutional and disciplinary settings – which provided different sets of options in responding to the discovery. Researchers came from three kinds of institutions: universities; industrial research labs; and non-university research organizations.

University researchers were the largest and most enthusiastic group (Felt, 1996). In Germany, two-thirds of the researchers participating in publicly-funded programs came from universities; the proportion was even greater in Switzerland; and university research dominated completely in Austria. Why were university scientists so prominent in the new field, despite policy's focus on applied research? What were their motives to participate so numerously?

Superconductivity has been called the "phenomenon of the century"; as one of our interview partners noted, "no other phenomenon of this century has brought Nobel Prizes for eight scientists." But superconductivity had also been declared a closed field. A theory was in place explaining most observed phenomena, and experimentalists regarded further work to reach higher critical temperatures as futile. So investigations had shifted from basic to applied research. With the Müller and Bednorz discovery, and even more with the yttrium superconductors developed by Chu and Wu, this view and the BCS theory were cast into overwhelming doubt. The BCS theory predicted a 35 K absolute upper limit to critical temperatures, and suddenly materials were exhibiting superconductivity at 90 K. For many scientists, this was an irresistible challenge to develop new, or modify old, theoretical models to account for the new classes of materials. Schrieffer wrote, "Each theorist brings his own set of techniques – and more importantly, prejudices – to the party, with most new ideas really being an outgrowth of past work" (Anderson & Schrieffer, 1991). Others wanted to achieve a higher quality of experimental results in order to better understand the new compounds.

The discovery of HTS had great symbolic value for many, especially for those who refused to believe that research should be planned decades in advance or that scientific advances can be steered by bureaucracies. Especially in Europe and among university researchers, the original discoverers' work was taken as a vindication of science that could be practiced by individual researchers on a small scale, rather than as part of a huge program or a predictable strategy (Nowotny, 1991). Another dream flared up briefly: that individuals – like Müller and Bednorz – could make a great contribution by persistently following their ideas.

Although HTS posed fundamental questions for basic research, it also suddenly revived older technological dreams of loss-free current, high-speed trains, and other "miracles". The prospect of commercial applications immediately raised hopes that industry and government would provide increased funding for university-based research. Many researchers jumped on the bandwagon for opportunistic reasons. When the focus of the programs became clearer and the initial euphoria made way for a more realistic assessment, many new entrants would drop out again. Applications, the market, and the enhancement of university–industry relations played a central role in establishing HTS, indicating that Europe's research system is in a phase of transition to a state in which basic research is dominated by considerations of use (Rip, 1990).

Finally, throughout the world, solid state physics had lost much of its attraction. In the 1980s, the number of researchers in the field had doubled while funding remained static. Scientific progress had also stagnated. One interviewee described the field of semiconductor and metal physics as "collecting stamps", since work had reached a phase in which investigations burrowed into detail without expecting any surprises. Another said he had believed that in "solid state physics everything was settled; there was still work for engineers, but the concepts were all there". The discovery of HTS and the influx of new funding thus gave fresh impetus to the entire field of solid state physics, which suddenly found itself back in "the wild years".[1]

University researchers enjoy the freedom to choose their research topics. They are generally less tied to specific research projects than are their colleagues in industry or "big science" labs, where projects are often planned years ahead and linked to large investments, and where success or failure are judged by the achievement of specific research and development (R&D) aims. Their only restriction is that they are expected to regularly publish results that are sufficiently innovative to pass peer review procedures. They have to make sure they can fulfill these requirements. Thus, to counterbalance the unpredictable nature of basic research, they generally follow several research lines at the same time, ensuring that they have alternatives if an avenue of research comes to a dead end. This strategy can only work for groups of a considerable size.

Not only are university researchers highly flexible, they also have a pool of young researchers at their disposal – Masters and PhD students. These are easily attracted to new research fields, allowing a rapid start. Students are a large, inexpensive source of manpower. However, our study shows clear limits to their use. A balance of experienced researchers and young students is a successful combination (as we witnessed on several occasions); but the idea that a lack of senior staff can be compensated by a great number of students proved to be a fallacy. Students require much coordination and supervision, contribute little experience, and often leave the group after a short period of collaboration, taking with them the tacit knowledge they have acquired.

What of the much-vaunted interdisciplinarity of university researchers in the case of HTS? Did disciplines differ in the kinds of research questions and methodologies they chose? Indeed, with the discovery of HTS, scientists experienced disciplinary boundaries breaking down in favor of more strongly question-oriented research. Although the first impression was that HTS research was equally open to all the neighboring disciplines, in reality the field was quite polarized in this first phase. Physicists, in particular solid state physicists, were the most numerous group to enter the field, followed by chemists (Jansen, 1991). The majority of those physicists were in possession of a particular skill, knowledge, or know-how they had gained in a quite different scientific context, and they wanted to apply it to HTS. A high degree of mobility was thus observable for those researchers specialized in particular investigation

---

[1] Funding fell far short of meeting the scientists' hopes, and materials research does not seem to have overcome its crisis. In a 1991 article in *Nature*, materials research was described as "a hot field threatened by an icebath of inadequate funding" (Anderson, 1991b), and a *Physics Today* article stated that "the only hope was the announcement of the White House Office of Science and Technology that a major initiative for advanced material research is being formulated for President Bush's fiscal 1993 budget" (Goodwin, 1991).

techniques. The second big group was those who had been working on LTS and who were thus thought to have valuable experience with the phenomenon. Chemists on the contrary are said to be more linked to materials themselves than to any technique investigating them, and thus to be less mobile within the research field. Collaboration between these different categories of scientists was on a one-project basis, limited in time, and generally based on a work-sharing model. Chemists expressed fear of becoming mere auxiliaries preparing samples for the team. And indeed, they were rarely fully integrated in physics groups at the cognitive level. This limit to cooperation between physicists and chemists has tradition in materials research, as some of our interviewees noted.

This calls for a comparison with the United States, where the field of materials science had been firmly established already in the 1960s. Physicists and chemists alike had come to see themselves as materials scientists, producing new materials and studying their physical properties. The new hybrid of chemists and physicists was institutionalized by the establishment of several government-funded materials science labs By contrast, Europe's materials science, never as robust as in the US, was stagnating by the 1980s. Most physicists interested in making their own materials had worked in the United States at some time. But chemistry had a strong tradition in European universities, and chemists pursued highly specialized careers divorced from physics. This separation led to the development of different traditions and methods of dealing with research problems, which now had to be confronted and brought together again.

In contrast to the universities' focus on basic research, industry's main concern in responding the new challenge was clearly the development of marketable products. But an understanding of the basic features of the new materials is a precondition for setting up and optimizing production procedures and for investigating the possibilities of new technological applications. When industry realized the fundamental character of Müller's and Bednorz's discovery and, even more so, of Chu's discovery some months later, its reaction was cautious encouragement or reserved optimism. Everyone saw that a phase of basic research was necessary, though perhaps not precisely what universities conceive of as basic research. Only basic research would provide a basis for the strategic decision on whether to invest or not. Two big sectors of industrial applications were seen as especially promising: microelectronics; and energy and magnets technology. Progress in the former was linked to advances in producing extremely thin layers of superconducting materials; the latter required the much more complex and messy development of polycrystalline bulk materials in wires of sufficient quality. The decision whether to continue industrial research was complicated by the insecurities of banking on a technology that did not yet exist. The materials developed so far showed poor current-carrying capacities, it was hard to say whether they could be produced in a form satisfactory for industrial use, and there was no telling what the future markets might be.

To what degree does infrastructure impinge on the organization and structure of industrial research groups?

In Germany and Switzerland, industrial research groups had finances and infrastructure as good or better than that of university groups. Once an enterprise decides to join a research field, it can invest flexibly, unbureaucratically, and rapidly, assuring a smooth start. Management deals with organizational questions, so industrial researchers do not have to spend time on administration, write research proposals "selling" their ideas to funding agencies, or supervise students. Also, industrial research teams are generally larger and their level of qualification higher than in university groups. Industry researchers are rarely graduate students who leave when their PhD is complete; they are employed in stable, reasonably paid, long-term positions. They thus acquire much tacit knowledge and are prepared to try new ways of working, which is sometimes difficult at universities. Decisions are not made by individual scientists or small groups, but rather by management – and not based on scientific criteria. Thus, industry researchers must be flexible, subordinating their own research interests to their assignment. In purely academic research, basic understanding is the goal and quality is the criterion for choices; industrial research's goal is to develop applications to be sold on the market.

Industrial research has its own drawbacks and difficulties. Typical for HTS was the need to bring together expertise from a range of disciplines in pursuit of common objectives. Academic researchers move freely in the huge complex of a university, or even several universities; affiliation with a given institute is no hindrance to collaborating with researchers from another. A company, on the other hand, must draw on outside know-how, either from universities or other science labs Hoechst, for example, had to set up a completely new knowledge base, since it entered the field from the chemistry side and had no experience with superconductivity. To balance this, it tied scientists to its research programs, in some cases with exclusive contracts.

Another factor making collaboration difficult for industry is the desire for patents. They are the key to the company's future – economically, rather than in terms of scientific prestige or recognition. Many researchers point to Chu's behavior to support the claim that the concern with trade secrets has entered the university, but we saw little evidence of it in the European university labs we visited. However, from the very beginning, Hoechst was eager to play in the world league of industrial companies by gaining as many patents as possible, even if this meant keeping research results secret.

The economic significance of the new field of HTS was overestimated in the first flurry of excitement. It will take many years before the new materials become essential for the markets for superconducting devices, and it is still very unclear whether and which new technological possibilities will open up. Nevertheless, industry's interest in future markets played an important role in getting national research efforts started.

In continental Europe, government-funded non-university research centers occupy the middle ground between universities and industrial research labs Funded on an institutional basis largely by public money, their research is organized differently than at the university, and they have their own distinct administrative structure. How did this influence their response to HTS, and what role did they play in the broader national context? In Austria, their contribution was negligible. In Germany, they played a very active role from the beginning, setting up their own initiatives and, even more importantly, acting as a catalyst on the national level. They carried out a significant part of the research effort, in accordance with the size and importance of government research centers there.

As widely recognized experts in conventional superconductivity, the Karlsruhe researchers led the effort to set up a national research program, including the university research. All three non-university institutes in Germany started much earlier than the joint university projects, partly due to the advantages of independent financing and stabler organizational structures. The picture is less clear in Switzerland.

In Germany, non-university research institutes have far superior financial resources than universities; their situation is more secure and they can plan well in advance. Researchers do not need to spend much time or effort on applications for additional funding from the *Bundesministerium für Forschung und Technologie* (BMFT) or industry; and if more funding is needed, it is management's and not the researchers' task to secure it. But long-term planning also limits autonomy in altering direction. Groups committed to other programs participated late or not at all. In this sense, university researchers purchase their greater freedom to choose the subjects of their research with the burden of finding the resources to do so.

Though university and non-university groups were of similar size, their structure differed fundamentally. Non-university research teams tend to have higher average qualifications, while university teams include many students. Non-university researchers also tend to have more research experience than university researchers, since they are free of teaching and administrative tasks and can do research full time. Finances and qualifications allow non-university labs to pursue several promising avenues at once, increasing the odds of success. Finally, basic research in non-university laboratories has a different meaning than at universities; it is meant to provide a base for pursuing the lab's mission.

Studying "big science" labs allowed us to observe basic research and applications-oriented teams at work in the same context; this heightened the contrast between the responses to the new research opportunity. The most important difference was in collaborative behavior. Measured by the number of scientific publications, in-house cooperation plays a much greater role than do links to the rest of the scientific community outside the lab This is especially true of the applications-oriented groups, far less so for basic researchers. The scientists seem to identify very strongly with their institution, which they sustain by producing competitive results. The institution matters to them almost as much as their own careers; university researchers are far more individualistic. But competition within the organization is high; this is particularly perceived by those involved in basic research. While applied researchers draw from the wide pool of basic research results and cooperate more readily, basic researchers employed avoidance strategies to delimit their "territory". This should be seen in the context of international competition, where awards and other academic recognitions are at stake, and where time presses more urgently than in applied research. But at non-university research centers, new developments or technologies are attributed more strongly to the institution and less to a single researcher.

3.2 Shaping the research environment

The excitement about the discovery of HTS arose in the context of a changing science system. Evident for some time, the signs of strain have been noted in the discussion about science in a steady state (Ziman, 1990, 1994; Cozzens *et al.*, 1990), in numerous committee reports, and in frustrated scientists' letters to the editors of scientific journals. In contrast to previous decades, funding for research is leveling off, a trend only exacerbated by the collapse of communism and the concomitant reorientation of military research. The implications of the present transformation of the science system are far-reaching. The foremost issues in the ongoing discussion include the increasing market-orientation supposedly permeating scientific practice and the attitudes of scientists, as well as the altered university–industry relationships that demand a new role for the universities. Growing interest in "strategic science", "priority setting", "selectivity" has become commonplace, and economic competitiveness and the applicability of research findings seem to have become central criteria for scientific choice. Researchers are thus compelled to develop mobilization strategies if they want to retain control over the kind of research they wish to pursue. However, the structural reasons for the widespread malaise have not been well articulated. Not all the changes are the simple result of exponential growth – more scientists, more funding needs, higher costs in research – reaching a saturation point where quantitative growth gives way to chaotic fluctuations until a qualitative change in the system occurs.

In the past, academic research has been the prime beneficiary of the rapid expansion of the science system. More recently, there has been a decisive shift. The production of knowledge has drawn on multiple resources across academic disciplines and across industrial sectors. It no longer restricts itself to academia or even the industrial basic research lab, but takes place in multiple sites. Users and producers of knowledge now work more closely together; their novel configurations reinforce the heterogeneous diffusion of knowledge production. The new multiple sites of knowledge production are also becoming more international. It is no longer possible for a few highly industrialized countries to hold a de facto monopoly over research results they have produced by themselves: today, technological and commercial benefits accrue to whoever is capable of utilizing the new knowledge. Scientific growth has become heterogeneous, generated by ad hoc configurations rather than stably structured hierarchies (Gibbons *et al.*, 1994).

While these changes have been recently observed in greater or lesser detail, buttressed by statistics and a flourishing literature on the economics of science and research, less systematic effort has been devoted to their theoretical understanding. Relatively little attention has been paid to the effects this extension of and these drastic changes in the research system have on researchers. For instance, pressure to publish has increased so tremendously for many researchers in the life sciences that fraud and ethical conduct are becoming important issues. Public discussion in the scientific press now devotes regular attention to the conflicts of interest of scientists who hold stock in pharmaceutical firms or of journals refereeing articles containing commercially sensitive information. Such issues also concern university boards and other organs engaged in revamping the rules of scientific conduct.

Scientists' lives and the scientific culture of the lab are also changing in many other, less dramatic but consequential ways. In the European laboratories we observed, where researchers had plunged enthusiastically into the new HTS field, their nostalgia for an older way of doing research was bound up with the large share of activities not properly termed research that preoccupied them. This was not merely administration, as opposed to "real science". It consisted of genuine intellectual and scientific preoccupations translated into and triggered by social interactions and involving scientific expertise, access to knowledge, organizational and financial resources, and scientific authority. These resources were invoked, manipulated, and socially negotiated to sustain claims that pursuing basic research would produce future scientific and technological benefits.

Activities other than research are conventionally viewed as alien to science, a regrettable waste of time and energy better spent in doing science properly speaking. But, as we saw throughout our study, these other activities have become an essential precondition for doing research at all. Science policymakers and lab directors used to reach a career turning point when they became administrators, setting up research programs, finding funds, and managing the scientific work of others. But all scientists now spend a considerable portion of their time attempting to shape their research environment.

In each lab we visited, we encountered more than the usual share of the administrative and teaching duties that form part of the normal life of European university scientists. Our interviewees told us about their concerns and about their strategies for overcoming obstacles to pursuing the kind of research that interested them. Practically all university scientists (and even more so the research managers and directors in non-university research settings) were intensely engaged in what Krohn and Küppers (1989) have called *Wissenschaftshandeln* ("scientizing"): implementing strategies on the local as well as the national and international level, setting up research groups or collaborations, obtaining information about research results and promising new directions – all with the goal of writing a research proposal and of taking part in a program that would provide them with funding. Granting interviews to the local press or setting up an information seminar for students was as much a tile in the mosaic of activity as was searching through the masses of new scientific literature, communicating by phone or e-mail with colleagues abroad, or using personal clout to persuade policy-makers in ministries, funding agencies, or industry that they, too, should help set up a national research program. These activities demanded a considerable investment of time and energy, of informed strategic planning, and of social interaction to find collaborators or to distance oneself from colleagues one did not wish to be associated with (a common feature of European university life).

To term all these activities "fund-raising" would be as absurd as to see actual research work as nothing but "spending research funds". The conventional view that research has a self-evident and autonomous status is inadequate. Our interest focused on the interface where research, its content, and the conditions under which it is carried out are negotiated. To this end, we adopt the distinction made by Krohn and Küppers between "researching" (*Forschungs-handeln*), defined as any activity directly linked to the pursuit of knowledge production by a research group, and "scientizing" (*Wissenschaftshandeln*), which includes all interaction with the wider environment. The latter structures the environment and creates the conditions necessary for research to proceed in its own autonomous and sometimes anarchical ways (Krohn & Küppers, 1989; 1990).

This use of the term "scientizing" to include activities other than research, does not, however, refer to activities pertaining to the organizational shell that are untouched by cognitive content, nor does it consist solely of the social and political skills that might benefit a good researcher. Rather, it permeates all activities designed to shape the environment in order to be conducive to research.

Among researchers who had entered the new field of HTS research, we encountered an astonishingly broad range of activities aimed at shaping the content of research programs and the directions into which research should move. The status of expert in this newly founded field became a central issue. As Limoges has pointed out convincingly, expertise cannot, however, be reduced to a set of information uttered by an expert. Expertise must rather be seen as a process "which in the end defines the status of knowledge and sets the limits of efficacy. This entails that expertise is not the property of a given individual, the expert." Rather, expert status is at stake in public forums – in our case, those of policy-makers, science administrators, and the media – and has to be re-established each time a major restructuration takes place (Limoges, 1993: 418).

While Limoges deduced this from his study of decision-making processes in controversies in the public domain, we find interesting parallels in the early phase of the establishment of HTS. Indeed, we have two groups of actors competing for influence on the policy level.

One group consisted of scientists with a background in LTS. They had either abandoned the field when basic research had stalled and funding dried up, or they were still involved in ongoing projects to develop applications or other marginal areas. Though the HTS ceramic oxides differed fundamentally from the metal compounds of LTS, members of this community were confident they possessed the required expertise to play a leading role in establishing the new field. Competition for leadership was rarely openly discussed, but it was clearly reflected in how our interviewees justified moving into the field: they were linked to this research "via a long tradition"; they had always been "in close touch with the phenomenon"; they could look back on "twenty years of activity in the field"; or the project was likely to be "problematic if no superconductivity specialist took part". One interviewee even said that, although the work of Müller and Bednorz was fundamental, they were mostly just lucky, since they did not really "know how to do research on superconductors". In the early phase of setting up HTS national programs, the veteran LTS specialists played an essential role in negotiating with governments and funding agencies. For a time, they were the only plausible experts in a field still lacking other qualification criteria.

The other group entering HTS research, the newcomers, were materials specialists, physicists, or chemists familiar with investigation methods that could be applied to the new materials. For the university scientists in this group, the most important question was whether their expertise and equipment could be effectively applied to HTS. But newcomers in applied research groups in industry or big science labs were more interested in developing commercial applications, such as high-field magnets for fusion reactors or Josephson junctions for microelectronic circuits. The chemists contributed expertise in making new compounds and controlling processing conditions; their methods and objectives complemented those of the physicists, making them valuable collaborative partners. The crystallographers offered knowledge of the basic molecular structures of the new materials. Other scientists, mainly physicists, were specialists in such sample preparation techniques as thin-film production using chemical vapor deposition (CVD) or sputtering. The new superconductors, whatever their form, would have to be precisely classified and their intrinsic properties analyzed, so a broad spectrum of experts in microscopy, spectroscopy, magnetism, etc., was also attracted to the field.

Relations between these two groups and the credibility of first research proposals varied in the countries we studied. In Austria, only two of the fifteen groups entering HTS had previously worked on conventional superconductivity. In Switzerland, with its long LTS tradition, a number of research programs on conventional superconductivity were still in progress when Müller and Bednorz discovered their oxides. In Germany, most of those initiating HTS research at a university had some connection to LTS, and the "big science" labs had carried out projects on superconducting high-field magnets. In the Netherlands too, a tradition of LTS had been kept going, which now was drawn upon.

Activities at the micro-level, alone or in collaboration, were echoed, magnified, transformed, and finally provided institutional robustness in a national HTS research program with at least a medium-term time frame and a budget to actually carry out research. In this process, research and activities directed at organizing the preconditions for research fed upon and mutually reinforced each other, overlapping and intersecting, but also creating a continuous conflict in the limited time budget of the researcher-scientists. These interactions took place everywhere from the micro-level of the research group to the meso-level of the national research system. There, activities to organize the preconditions for research widened to include policy-makers in ministries and funding agencies, decision-makers in industry, and politicians. All of these figures interacted with the media on occasion.

An even wider circle was reached on the macro-level of worldwide economic competition. The vast meshing between research and creating its preconditions is a measure of the complexity of the social construction of science and technology. The elaborateness of policy-constructs embodies society's economic aspirations, which constantly interact with the vast range of scientific and technological possibilities continually opened up by research. Economic and technological considerations are most visibly and rhetorically invoked on the macro-level, but they percolate throughout the system. They make themselves felt in the activities of individual researchers and research groups as well as in the decisions of research councils and the feasibility studies guiding industry. Starting conditions and degrees of preparedness to exploit novel opportunities differ enormously, partly as a result of previous research and organizational work, which is embodied in the institutional shape of a national research system.

The initial excitement about HTS and its aftermath were the manifestation of spontaneous collusion among the participating actors. In the first phase of discovery, HTS was almost unanimously described as a unique and splendid opportunity to research and to organize conditions for future research. In the second and third phases, these common interests were slowly and patiently transformed into more or less coherent sets of national research policies sustained by unswerving belief in the technological potential of the field. Research and strategic efforts to shape the research environment needed a third "leg" if they were to succeed in setting up a new and viable field in a very short time: the clinching attraction and ultimate legitimation were based in the credible claim that research on the newly discovered superconductors would ultimately yield technological benefits to society. A set of beliefs and activities was constructed and nourished that was oriented toward achieving the technological potential thought to inhere in scientific research. These beliefs and activities in turn extend into and structure basic research, even though no one can know if the technological hopes will "pan out".

Students of technology have insisted that it is impossible – whether in the initial design, development, and diffusion of a technology or in its mature state – to clearly distinguish between phases or activities that are technical or scientific and those guided by economic, social, or political consideration (Bijker *et al.*, 1989; Callon, 1989: 84). Thomas Hughes has convincingly demonstrated that technological systems:

.... contain messy, complex problem-solving components. They are both socially constructed and society-shaping. Among the components in technological systems are physical artifacts.... they.... include organizations.... and they incorporate components usually labeled scientific

(Hughes, 1989: 51).

In the case of HTS, the heterogeneity and complexity of technological innovation was manifest from the beginning. The emergence of the new field reveals the dynamic and dense intertwining of the three activities – research, efforts to organize research, and creating technological expectations. The "bright spot" of the unexpected discovery became the seed of a new field. For the countries we studied, we present the empirical detail of the social, organizational, and cognitive mechanisms that were required for this seed to grow.

An unexpected event triggering the growth of a research field displays the contingency inherent in history. In hindsight, the event seems to have occurred at just the right moment – but what context defined this moment? Many researchers saw the new discovery as a vindication of "little science", the kind of research open to everyone, even those without huge budgets. While this notion soon proved an insufficient basis for continuing, in the meantime it had helped mobilize and motivate many researchers. On the level of the preconditions for research, HTS functioned as a test case for existing national science and technology policies. It was seen as a catalyst for organizing national research programs along changing policy lines. Here, too, collusion prevailed. Scientists are well aware that economic promise is now one of the decisive criteria for basic research funding. They invoked potential technological benefits and national economic competitiveness in efforts to win governments and industry as allies in national research programs. Though empirical evidence shows that the link between basic research and technological performance is tenuous, the argument was repeatedly advanced that scientific competitiveness forms the basis of a nation's technological economic competitiveness. And national policy in many countries adopted the same argument – including the bugaboo that others might be faster. Thus even competition between nations functioned as objective collusion to put the field on its feet.

3.3 Communication, cooperation, and competition

Among the intangible components that shape this new research territory are social interactions, including communication, cooperation, and competition between individuals, research groups, and national science systems. Personal elements, structural and institutional factors reflecting changes in the science system, a country's research traditions, and the flexibility of its response to new developments are all important in determining the framework for these interactions and the impact they have.

With the general recognition of these social interactions' importance for the science system, increasing effort has gone into collecting data on scientists' communicative and cooperative behavior, including the number of co-authored papers and their distribution among various disciplines and geo-graphical regions, researcher mobility, new communications media, and the increasing density of cooperative networks on the international level. These often quantitative results then play a role in policy deliberations. But they are only the tip of the iceberg, tending to neglect the multi-layered, informal, and often not easily discernible networks that are the very basis of the science system. HTS provided a powerful reminder that these informal interactions create and exchange crucial "symbolic capital" that can be used to negotiate financial and structural support for the budding field. Mere numbers of exchanges are a very crude measurement of the complex qualitative phenomena of communication and cooperation.

Most scholars in social studies of science would agree that competition is an important spur to scientific progress. But this phenomenon, too, has not been subjected to much detailed study; in particular, its impact on knowledge production is not well understood. HTS research resulted in competition on several levels: between individual scientists, between research groups and institutes, and between national economies. These levels of competition can probably be observed in many fields on the border between basic and applied research, but HTS had another level of complexity: the media offered a public arena and an expanded set of criteria for success.

Communication is at the center of any academic enterprise. The promotion of knowledge (the main cognitive concern) and the establishment of reputation (the key social concern) are necessarily dependent on it – Becher noted, and – it is the force that binds together the sociological and the epistemological, giving shape and substance to the links between knowledge forms and knowledge communities

(Becher, 1989: 77).

HTS provides a good illustration of the impact of communication on the process of organizing a new research field. The production and distribution of scientific and technological knowledge has accelerated unprecedentedly. Although we have grown accustomed to scientists communicating by phone, fax, and e-mail to exchange new information, in the months following the HTS breakthrough, we witnessed not only a much higher volume of exchange, but also a change in the status of the communicated information. Scientific results were publicized at a much earlier stage in the knowledge production process than ever before. One of our interview partners made this explicit when he said that, in the first phase of research, the greatest expenses were incurred not in doing the research but in paying the telephone bills. At a time when thousands of researchers were at work in this field and new HTS compounds were announced almost every week, it seemed essential to communicate new ideas and findings quickly, in order to document priority. Publication delays of a few months seemed intolerable. Results were often published before they were verified and in terms too general to permit verification by other members of the scientific community. In this highly competitive situation, the speed of the exchange of information began to dictate the speed of what appeared to be scientific progress itself.

Aside from speed, the multiplication and differentiation of information pathways were salient features in the months following the HTS breakthrough. Very rapidly, electronic bulletin boards were established, unrefereed journals and commercial newsletters were founded, and the media were used to disseminate scientific news. The relative importance of various communication channels within the scientific community also seemed to be renegotiated for a few months. Normally, a research field's knowledge base consists of stable information in the form of articles refereed before publication; in HTS, for awhile this role was assumed by the temporary, far less stable knowledge of electronic bulletin boards, faxes sent around the globe, hastily printed newsletters, or even daily newspapers. "From faxes to facts" (Lewenstein, 1995a) and "science by press conference" were two ways scientists tried to disseminate new results and get them accepted by the scientific community. This triggered a broad discussion among the scientific community, policy-makers, the public, and in social studies of science on the changing "norms" of the science system. The analyses:

Especially problematic is the degree to which the new electronic technologies can replace traditional face-to-face interaction – the informal conversations, presentations, and discussions that give shape to the more formal aspects of scientific communication such as conference abstracts, refereed journal articles, and textbooks

(Lewenstein, 1995b).

Apparently, the traditional publication mechanisms had been pushed to their limit and, at least for a short while, were being replaced by other information channels. HTS might be a good case to judge the impact of this new communication behavior in a few years time.

Publication is still considered an important product of scientific research; but the phase of information transfer has been shifted to a much earlier time. It could be argued that this is nothing fundamentally new, since exchanging preprints has long been a common ritual in the scientific community and many scientists consider it important to be on the mailing lists of the major research centers. But here the exchange was even one step earlier, on a more informal level.

The case of HTS also clearly underscores that communication has become virtually simultaneous all over the world. Even in very peripheral locations, information networks make more and more knowledge available, prior to publication in conventional journals; membership in such networks has become crucial for those wanting to be at the cutting edge in the field. To have been in a place, attended a meeting, been personally informed by a colleague is coming to play a greater role than reading the published result, if one wants to be able to respond quickly to changes on the forefront of science. Thus the researchers' ability to move from one site of knowledge production to another has become a crucial factor in the science system. Physical presence is the most efficient way to exchange ideas and negotiate their degree of realizability, as well as to exchange tacit knowledge.

Membership in such informal networks means access to restricted information that often exists only in verbal form. Traweek pointed this out in her study of Japanese and US high-energy physics communities (Traweek, 1988). She not only hinted at the centrality of verbal communication in the science system, but also showed how multi-faceted the impacts and consequences of these informal structures are, on both the social and the cognitive level. One talks with colleagues to evaluate one's own work and ideas, to seek allies, to persuade others of the importance of one's work, to judge the progress of one's communication partner and of other close colleagues, or to prenegotiate what will later be written in applications for funding. Oral communication is thus a powerful means of including or excluding colleagues, of exchanging "symbolic capital", and of creating boundaries within and around a community – which is especially crucial for a community in the making.

In most countries, the first initiatives were mounted by individuals and the first negotiations were conducted by telephone or in small, usually informal meetings. Good international and national personal connections in the field were an important asset.

This is nicely illustrated by several scientists' personal accounts of their relationships to other parts of the science system. HTS was discovered in a multinational corporation's lab in Switzerland by a Swiss and a German and they published their results in a German scientific journal. But the wave of excitement – and the flow of information – that spread through the scientific community emanated primarily from the United States. The key factor in establishing the credibility of the Müller and Bednorz findings was their replication by US researchers. Our interviews showed that the first researchers to respond to the discovery either had excellent direct contacts with a US lab where they had once worked or where their former students now worked or else happened to have been in the US when the first announcements came in late 1986. These European scientists carried the news back to their home institutions. Public presentations at universities, preprints, talks at conferences, telephone calls, faxes, and e-mail messages assured rapid diffusion. Rapidly-founded regular news bulletins reported successes in achieving superconductivity at higher temperatures with new materials developed by the growing number of teams entering the field. After the March 1987 meeting of the American Physical Society, detailed and regular reports in the US media gave HTS research high public visibility and enormous prestige, and saw in it the prospect of lucrative applications in the near future.

Our case studies show that, for the European scientific community, the United States often functioned as a decentralized but well-connected communication center. The fact that the collaboration network, measured by joint authorship listed in the *Science Citation Index*, was centered in the United States (Schubert & Braun, 1990) probably also holds for HTS. The US also influenced the shaping of research traditions, the idea of scientific leadership, and the sharing of tacit knowledge. Many of the scientists who had constituted the core of European LTS research had spent decisive periods of their careers in a US lab Some had been students of an outstanding American scientist like Bernd Matthias, or had maintained good contacts with US labs One Swiss interviewee, for example, described setting up his institute in Switzerland as "a strange genesis, which drew its material out of this huge [US] laboratory, but which had traveled here mentally and had become implanted". The experience he gained in the United States was the basis for his own vision of what a successful institute should be, and he realized this vision. Similarly, a German scientist said he got the idea of how to set up his team from his stay in a lab in the United States.

The importance of the "American connection" varies considerably within Europe and between disciplines. Traditionally, many Swiss and German researchers have strong ties to the United States – our sample shows this to be particularly true for physicists, less so for chemists. They believe this has been important for their intellectual development and for how they organize their work. In Austria, such ties are rare exceptions. The last OECD (Organisation for Economic Cooperation and Development) report on Austria maintained that its science system "suffers from what it feels to be a kind of isolation" (OECD, 1987) – a conclusion confirmed by our study. Austrian researchers tend to be less mobile than their foreign counterparts. Despite a florid rhetoric of international cooperation, it was regarded as less important when it came down to making decisions and what was happening in the international science system weighed less in Austria than in other European countries. Many Austrian scientists doubted that their national system could maintain European or especially US standards, anyway.

What was the relation of group structure to communication behavior within the research groups and toward other groups on the national and international level? In those groups with a large core of experienced scientists plus a number of PhD students, communication was distributed equitably among all group members. Most got the chance to attend major international conferences to present the team's scientific results, and thus to exchange ideas on the international level, take part in coffee break gossip, get an idea of other people's plans, gain an overview of what was going on in the field, and see their own work in the context of what others presented. Since many researchers linked the group to the outside at the same time, a variety of views and perspectives was fed into such a group. No clear selection took place on this level.

The other extreme was groups with a single leader at the center of an inexperienced, mainly undergraduate staff. Here, group composition fluctuated and much tacit knowledge was lost when students left after finishing their course of study. Those leaving had no direct communication with the new students joining the team. Especially in some Austrian and German teams, much time had to be spent on regular communication within the team to assure coordination. Contact with the outside was channeled almost exclusively through the team leader, a clear if unintentional information filter.

The structure of the communication system and the prominence and speed of the channels for dissemination played an important role in competition, which was particularly intense immediately following the discovery, as is evidenced by media accounts, the temporary breakdown of the peer review system, and the launching of HTS newsletters and journals. The media's simplified story of this competition differs from what actually happened among the various research groups. Part of HTS's attractiveness for the media, especially at the outset, was that competition could be translated into a race for higher critical temperatures. All the ingredients of a sporting event were present. There were individual players with easily measured performance and clear rankings, based on the timing of their achievements. Temperature charts were an early feature in almost all print media. Later, the US media in particular described national research as a "sports team" competing with other national teams, especially Japan's.

The social reality of competition among the researchers was somewhat different. Open, race-like competition with a single victor is the exception, not the rule. "While there would be general agreement that science is essentially a 'competitive world', it is clearly not the case that all scientists are engaged, all the time, in open competition with each other" (Edge, 1990: 231). Mechanisms prevent the open expression of competition.

The first few months of HTS research did resemble a race. They were characterized by a strong belief that anybody could join, at low risk, because apparatus to make first samples were "cheap, unsophisticated and easily available" – similar to the situation described in Gilbert's study on radio astronomy (Gilbert, 1977). Competition was not yet for funds, but for the "old currency" of science: recognition of merit. In the long run, such a restricted view of competition could not prevail as the explanation for what happened in the scientific community. The situation had changed when it became clear that Müller's and Bednorz's discovery had only been a start and that funding bodies would be willing to help building up a new research field.

This brings us to Latour's and Woolgar's notion of "credibility". They contrast the notion of credit as recognition of merit with the notion of credit as credibility, i.e. as personal influence based on the confidence of others or as a reputation for solvency and probity in business. Credit is thus associated with "belief, power, and business activity" and it has to be understood as having a much broader sense than simple reference to reward (Latour & Woolgar, 1986: 194). This broader sense of credit is captured by the term "credibility". Thus the scientists main aim was to "gain credibility which enabled reinvestment and the further gain of credibility .... In this sense scientists' credibility is similar to a cycle of capital investment." Being accepted as an expert in a domain gives privileged access to grants which enables one to improve one's equipment. This in turn leads to higher data quality, which helps to hone argumentation, which is in turn published and widely read in the community. This again reinforces expert status. Such a cycle of credibility, which contains both economic and epistemological elements, thus allows conversion "between money, data, prestige, credentials, problem areas, arguments, paper and so on" (Latour & Woolgar, 1986: 200).

Belonging to the LTS community, having worked with one of the leading figures in the field, or good links to industry in market sectors relevant to the field were all definite assets in the early phase of HTS. Experience with the funding bodies which would support HTS research or good international connections could be used as capital in negotiating one's position in the national programs. Holding a central position in a research field, enjoying expert status, or being an editor of a journal also increased one's influence on the perception of which research issues were important. In the case of HTS, this latter point was even discussed in newspapers. So we find the *New York Times* of April 5, 1987 quoting a scientist: "Only a handful of 'guru' editors and referees dictate what gets published . . .. Physics is what the 'in crowd' defines it to be. ... The criteria for what is acceptable are basically controlled by a very small number of self-proclaimed hot shots." And "as with money capital, the size and speed of conversion is the major criterion by which the efficiency of an operation is established" (Latour & Woolgar, 1986: 201) and by which competitive behavior can be optimized.

The third facet important in configuring HTS as a research field was the development of cooperative structures, nationally and internationally. The early phase following the breakthrough was clearly dominated by individual initiatives. Researchers worked on their own or in their existing groups. Thus the history of individuals and groups made a crucial difference, and early options for action in the field were dominated by previously existing networks. Personal connections and infrastructure were both important in seeking partners or in being sought as a partner.

Following this phase of unstructured, individually organized collaboration, and by the time HTS had entered the official policy discourse, scientists thought collaboration should be given more structure, especially in the context of global competition. Cooperation networks organized in the form of national programs would become the basic principle of order. Each country had its own ideas of how conditions for effective cooperation and exchange should be set up on the national level.

Generally, small countries can best enhance cooperation among university researchers through the subtle guidance of a steering group. This was surely the case in Switzerland, which has a long tradition of good cooperation within the country as well as with the international scientific community. Here, no formal rules of cooperation were tied to the allocation of resources. Austria, by contrast, attempted to remedy its lack of cooperation (regarded as one of its main weaknesses) by ordering it from above: "No cooperation, no money." Austrian scientists entered strategic alliances perforce, but it seldom led to the desired success.

Larger countries need stronger institutional mechanisms. In Germany, the BMFT regularly relies on mediating bodies such as the *Verein Deutscher Ingenieure* (VDI, Association of German Engineers) to ensure the implementation of cooperative projects between industry and university. As in the case of HTS, this may entail the informal selection of which university research groups to include, and sometimes even the stipulation of partners. Such management practices presuppose excellent contact with scientists and industry as well as comprehensive knowledge of their respective technical infrastructure and specific capabilities.

In Germany, the concept of locality was also rediscovered. The BMFT did not set up a national cooperative program, but asked various university institutes, industry, and non-university labs to form local cooperation clusters. University clusters included several disciplines, mainly chemistry, crystallography, and solid state physics. This not only assured smooth and active knowledge transfer across disciplinary and institutional boundaries. Attention was also paid to the fact that the transfer of knowledge and skills is tied to persons, and thus that geographical proximity plays a central role (Pavitt, 1991). Being able to meet informally, visit each others' labs, discuss ideas continuously, and supervise students all guarantee a better flow of ideas and lead to innovations.

3.4 Everyday life in the laboratory: the social reality of research

In the following, we want to examine the conditions of everyday life in the laboratory: the problems faced by the majority of the researchers at the workbench and the difficulties of transferring policy decisions into the routine context with all its constraints. This picture contrasts starkly with the fascination of the extraordinary associated with names like Müller and Bednorz, Chu, Wu, Tanaka, and others whose scientific feats were widely publicized.

Although we visited too few laboratories to permit broad generalizations, our observations and the impressions gathered in many informal conversations with the researchers do highlight some facets of everyday laboratory life in HTS research in the early phase. Our first contacts with the scientific community and the newspaper reports created an image of a field suffused with excitement, where scientists rushed to make new discoveries, their research requiring little investment. A vast rhetoric had been constructed about possible technological applications, future markets, and the economic competitiveness investment would assure. Science policy-makers saw HTS as a test case for their policy visions. Some tried to enforce well-meaning programs which stipulated work-sharing and channeled competition from the national level to the outside; others gave researchers more freedom to organize their research environment.

We observed three labs for one week each in Cologne, Germany; Graz, Austria; and Geneva, Switzerland. The realities we encountered had little to do with the policy debates or the rhetoric sustained by almost all actors involved. Of course there had been a great deal of excitement at the beginning, when many scientists were eager to join the research effort, but now, more than two years later, "normal" research conditions had returned. The high speed of developments so characteristic of this new field during the first few months had slowed considerably, and visions of the future had become more sober. HTS had taken a place on the research agenda side-by-side with other topics. Money had been allocated, often less than initially hoped for. It allowed researchers to join this fascinating new field, and it also guaranteed a better financial situation for awhile. But few groups moved entirely to the field.

Many glorifying accounts overlooked that much of the research needed to understand the new phenomenon was routine, repetitious, and sometimes boring. Making well-characterized, high-quality samples is a complex, time-consuming task bearing no relation to the media's early descriptions. It is not "shake and bake" (Maranto, 1987: 27), as one article quipped, not simply a matter of mixing ingredients to get "something" with a high critical temperature; rather, it means devising systematic methods of producing the new superconductors in various forms and with high quality, especially with a view to future applications. The amount of drudgery was increased wherever a lab was not well equipped with basic auxiliary machines. During our visit to Graz, for example, we were told that, for lack of a diamond saw, slices of the baked superconductor pellets had to be cut with an ordinary fretsaw; unsurprisingly, this led to imprecise and frequently unsatisfactory results. Thus, additional money was needed not only for special equipment, but in many cases just to support normal working conditions.

Even labs with good basic infrastructure needed new equipment to join the research effort. Preparing and investigating the properties of superconducting bulk samples, crystals, and thin films requires specialized and sometimes expensive equipment – a SQUID, for example. Groups wishing to remain at the forefront of the field over the long haul had to exert great effort and present well-argued funding proposals to convince funding agencies and governments of the need to finance both the necessary equipment and the qualified staff to operate it. This profoundly affected university labs in Austria (and in some cases in Germany), where resources were inadequate and additional money was necessary to conduct research at all.

Researchers confronted with the reality of meager funding often tried to adapt existing apparatus so that they could take part in HTS research. This often meant enormous investments of time and demanded a high degree of inventiveness and technical skills, while often leading to less than optimal solutions. Some measurements could be made, but without the degree of precision that would allow international competition. The students who usually did this work often felt they were wasting their time not "doing real science" and that their findings had no real significance. One student expressed his dissatisfaction very clearly: he said the solution was to leave the university as soon as possible after finishing one's degree, since research conditions would be far better in non-university labs and time and energy would reap better returns there. The lack of academic jobs meant most research students would have to leave after their degree, anyway. This meant the team would lose someone who knew how to use the equipment and would have to train someone new.

In the German and Austrian labs we visited, the percentage of students in the research groups was high. In Cologne, fifteen masters students and fifteen doctoral students worked on a team that included only one professor and two senior staff supervisors.

We originally assumed it was the spectacular breakthrough that had drawn the students to the HTS effort, but most of those we spoke with said their primary reasons were the person of the team leader and the working conditions he could offer them. Next most frequently, they mentioned egalitarian treatment and a good cooperative climate. The attraction of the research potential of HTS was only the third-ranked reason. Most students realized they would not be able to stay within the university framework after their degree, and this discouraged their "bonding" to any particular scientific question. While many accounts stress that students were a very important human resource, allowing the universities to respond flexibly to new research possibilities, this euphoria was dampened by the reality of the infrastructure. The HTS groups at Graz and Cologne suffered a great lack of space. This meant students were in the institute only while actually conducting experimental work and had to return home in the intervals. This isolated them from continuous communication. Regular meetings were held to compensate for this and assure a minimum of coordination, but they could not replace the informal contacts essential in the knowledge-production process.

The groups collaborating in Geneva were in a different situation. First, they had a large core of internationally reputed scientists with enough clout to convince the funding agency and who also served as role models for the younger researchers. Second, the number of students in the Physics Department was relatively small, reducing researchers' teaching loads at the same time as providing students with much better supervision. This helped the well-reputed postgraduate department attract highly qualified students. Most administrative tasks were dealt with on the highest hierarchical level, giving other team members (in particular younger ones) more time to do research.

Interdisciplinarity became a buzzword on the policy level. How was it realized in everyday research? After Müller and Bednorz demonstrated that complete outsiders could contribute so essentially to understanding superconductivity, researchers from many disciplines were attracted and felt free to try their hand in the field. But our small sample suggests that research reality was far less interdisciplinary than the rhetoric developed at the policy level implies. Most researchers involved in HTS were physicists, though new configurations were evident; crystallographers joined, and chemistry know-how was incorporated either by collaborating with a group from the university's Chemistry Department or – in the majority of cases – by employing a chemist to prepare HTS materials. Interdisciplinary links often already existed, as for example at Cologne, where collaboration with the Chemistry Department had a long tradition; sometimes they were intensified when strategic alliances were needed, but we did not see the breakthrough policy-makers sought in the mode of research – the durable establishment of new configurations and interdisciplinary cooperation on the conceptual level. Far more frequently we saw work-sharing, where researchers from different disciplinary backgrounds followed differing research questions in the organizational framework of one project.

The experience we gained from our visits to the labs confirms what Latour argued in his lab study. "What makes a laboratory difficult to understand is not what is presently going on in it, but what has been going on in it and in other labs" (Latour, 1987: 91). Indeed, researchers' abilities to respond to this unexpected breakthrough depend on more than their individual abilities to quickly grasp the scientific relevance of new findings. Researchers are greatly limited by the infrastructure prevailing in their laboratories, the work loads they face in teaching and administration, their existing research commitments, and the quality of their links to other labs meeting international standards. And these conditions differed enormously from lab to lab and from country to country. The contexts in which researchers were embedded dictated the choices they could make and the kind of research they could propose in the important initial phase. Compared to a particle accelerator, HTS is a case of "little science", but researchers still needed the necessary equipment. No country allocated money immediately; this meant that new devices could not be purchased for several months. Setting them up and hiring new staff took additional time. Only those lucky enough to have the right equipment from the outset could take the lead in research. Müller and Bednorz did not need expensive equipment to discover HTS, but the next steps toward understanding and exploiting the phenomena were another question.

The decision was not merely whether to join, but also how much time could be invested. The more an institute's money came from funding agencies, tying it to specific research questions, and the greater the number of staff members financed externally, the less flexibly could an institution respond. Most of the researchers we interviewed stressed their part-time involvement in HTS, for reasons of prior commitment but also for reasons of "risk management". HTS seemed promising, but no one wanted to put all their eggs in this one basket. But we have seen that, for poorly equipped teams, diversifying research to minimize risk backfires: spreading themselves too thin, they lose any chance of doing research on the forefront.

4

Academic research, science policy, and the industrial connection: setting up national high-temperature superconductivity programs

Science policy is rarely considered part of the core of the science system, but is seen as peripheral, providing a framework for establishing priorities and allocating research funds, setting up institutional structures within which research programs can proceed with a continuous, predictable level of funding, and providing incentives for the wider transfer of knowledge and for the utilization of research results – usually for the benefit of the national economy. The various models of science policy are closely related to the profiles of national institutions and to the specific instruments of policy-making available to obtain their objectives.

Many observers agree that science policy in the highly industrialized countries has moved through distinct phases or "eras" since the end of World War II. These phases differ in the underlying patterns of scientific and technological change, in the issues on research agendas, in the preferred instruments for decision-making, in the modes of funding, and in the modes of research. It is also said that, in the use and regulation of science as a source of strategic opportunities, science policy is undergoing a process of internationalization, in that international cooperation is being promoted (Ruivo, 1994).

Other science policy analysts diagnose an increasing "denationalization of science", evidenced by growth of trans-national research cooperation, even in less cost-intensive areas, the shift from public to private funding, and the regionalization of research (Crawford et al., 1992). But international cooperation on one level does not necessarily preclude competition on another. The globalization of the economy, the wider geographical distribution of the sources of scientific and technological knowledge, and growing interdependencies make it clear that the configurations of cooperation and competition are not fixed, but fluid.

Despite the trend to international collaboration and some convergence of the science policies of various countries, science policy is still predominantly conceived, oriented, and carried out within a national framework. Scientists have learned to live with this and put on different hats. In basic research, their professed outlook and value system is thoroughly international, in line with the much-cited universalism of science that calls for the free, unhindered circulation of ideas and people. But when it comes to research funding, scientists know they must first turn to their own national science system, with its array of funding agencies, programs, and institutions. Scientific careers involve geographic mobility, but they are still pursued mostly within a national context. To mention but one example, universities, which generally play a dominant role in basic research, develop in the context of and are guided by policies which remain largely a national affair.

Discussing the future of research in Europe, one of the organizers of a recent conference, "Research Policies for Europe's Future", claims it is not chauvinistic to channel the social functions of science – improving health, increasing wealth, and deepening general wisdom – back to the nation or group of nations making the decisive investments in science and research (Maddox, 1995). Europe's research community and decision-makers at the national and European level are seeking to enlarge the scope for research in individual European countries as well as on the level of European Union (EU) cooperation – and to find a viable balance between them. But the tension between the national policy level and the European level remains.[1]

Science policy exerts much more than a trivial influence on how and what research is conducted, and thus on the outcome. Institutions and policies channel funding over extended periods of time, set priorities, provide for prolonged and often stable links between academia and industry, establish working conditions favorable to creativity, and maintain or fail to maintain research infrastructure and technical equipment. They also powerfully signal to the next generation of researchers which career and funding opportunities are promising. Science policy infuses the scientific process with expectations of tangible returns, however loosely formulated or even merely implicit these may be.

HTS provides a good illustration of the unashamed, almost "old-fashioned" national orientation of science policy. The unexpectedness of the discovery, the intense, informal cooperation across national borders in the initial phase of research, and the alteration and even circumvention of traditional procedures of publication and funding were perceived as signals of change. What did not change at all was each country's basic assumption that it needed its own strong

---

[1] Of course, this tension is visible not only in research policies, but also in the reforms of the higher education system taking place on both the national and the international level.

national research program to enhance the nation's technoscientific and economic competitive position.

How did the various science policy systems cope with the discovery of HTS and its emergence as a new research field? What were the constraints and opportunities that determined whether an HTS program was successfully launched or ground to a halt after the first elated phase? Science policy systems differed in their preparedness, speed of response, and willingness to circumvent standard bureaucratic procedures when faced with the unexpected opportunities. Such readiness is connected with projections of future gains resulting from policy and with industry's willingness to contribute to the research effort, which often plays a crucial role in whether basic research is funded and gets off the ground. Also important are the historically-determined patterns of the university–industry–government triad, the activity or "restrained neutrality" of funding bodies, the clarity of the expectations placed on researchers, and the ability of the research community to organize itself. A well-organized community exercises more political clout than a dispersed, faction-ridden one.

National research programs are nothing new. They are established to develop fields assigned high priority by governments and research communities, often in view of potential applications and to strengthen national economic competitiveness in international markets. Most European countries have set up a variety of national programs in such areas as biotechnology, communications and information technology, and materials research. The European Commission follows the principle of subsidiarity (i.e. providing funds only in areas where the individual member-states cannot efficiently pursue research on their own) in supporting collaboration between EU member-states in academic research as well as in applied research with strong industry participation. The United States' federal funding system is geared to achieve specific national goals. In practice, the distinction between basic and applied research is rarely important as a basis for the allocation of funds by research agencies or Congress; they are subsumed under the term "precompetitive".

To achieve the aim of establishing national priority programs, researchers and policy-makers formed alliances we term "hybrid communities". These are no longer communities in the Mertonian sense of a set of actors, relatively stable in time, relatively homogeneous in outlook, and subscribing to the same set of norms. Rather, hybrid communities are heterogeneous, open, and fluctuating. Sometimes their set of actors is not easily discernible, since the individuals' roles, histories, and experience overlap and some individuals assume different functions in different positions within a limited period. These hybrid communities negotiate the parameters of policy decisions that shape the further development of the new field.

At the outset, HTS was unusually open, as demonstrated by the contribution of Müller, who disregarded disciplinary boundaries. With widely varying motivations, researchers from varying institutional contexts and disciplinary backgrounds, policy-makers in government offices, in funding agencies, and in intermediate bodies, as well as industry all rallied to foster HTS in their national research programs. The media, politicians, and the lay public were the spectators exhorting them, while representing the national interest.

The complexity and interrelatedness of the various components of science policy are apparent in the light of what science policy meant to these hybrid communities, what their central research questions were, what guiding concepts and policy constructs they used, and which established or emerging lines their research intended to follow. Quality criteria had to be defined and quality-control mechanisms established in a field where they could not yet exist, and in which there were many disciplines with equal claims to expertise.

Science policy had to decide how much money to invest in the field, and for how long. How should uncertain but plausible prospects for technological applications be assessed? Why did policy-makers believe the implicit and explicit promises made in each of the submitted research proposals?

In what follows, we will first take a look at some of the major industrialized countries (USA, Japan, the UK, and Germany) and the national research efforts they established. We try to analyze their starting positions, the motivations driving them, the policies they tried to follow, who got involved and what final form these programs took. In Section 4.2, we then turn to three of the smaller European countries (the Netherlands, Switzerland, and Austria) to see what policy options they chose in their search for a and find a niche where they could contribute to the new field of HTS research. In the concluding section, we try to analyze the importance of national policies, but also to show their contingency on numerous other factors in the science system.

4.1 Playing in the world league: USA, Japan, the UK, and Germany

USA vs Japan

National research and development programs reflect national scientific and political cultures, although the intent behind these efforts is to increase the country's economic competitiveness. As the economic stakes rise, so does the level of aggressive rhetoric on science and technology policy. President Reagan's HTS initiative, announced in July 1987, was part of the continuing rivalry, especially with Japan, for leadership in world industrial markets. Fear was expressed that the commercial exploitation of HTS, a future key technology, would follow the pattern of microelectronics in the 1970s. In that field, the essential breakthroughs – the discovery of the transistor and of semiconductors as well as the development of computers – had come in the United States, but the US share in the world markets for these products had fallen to 40%. The lion's share had gone to Japan; in some high-tech products, such as integrated circuits, the Japanese now controlled some 75% of world trade (OTA, 1988).

The image of technology as a mirage – a seductive but unfulfillable promise of solutions to all our problems – is compelling in the case of HTS. Predictions of levitated trains, endless supplies of cheap electricity produced by superconducting generators, loss-free transmission lines, and superconducting supercomputers fired the public imagination and led to the belief that the right policies would enable basic research to generate marketable products in a very short time. This belief was most fervent in the United States, where political speeches, newspaper articles, and scholarly books expounded how a still non-existent technology could give the US an advantage against its major rival, Japan.[2]

Debate has raged for many years about the differences between the United States and Japan in regard to funding sources, investment strategies, the roles played by the research establishments (including the US Department of Defense and Japan's Ministry of International Trade and Industry), the ability to link basic research to commercial applications, risk-taking, enterprise structure and culture, and Japan's long-term approach to technology policy. Japanese policies toward HTS as a strategically important research area looked familiar:

In case after case – from videocassette recorders (VCR) to multivalve cylinder heads for automobile engines – Japanese industry has made heavy (and often risky) investments in new technology on the basis of its ultimate promise, whereas US firms have held back, waiting for it to be cost-effective. That willingness to plunge in, combined with excellence in design and manufacturing, has allowed Japan to dominate technologies such as VCRs from the start and to capture others [such as semiconductor memory chips] even after US firms held a sizable lead

(*Robin et al., 1988: 42*).

The MIT Commission on Industrial Productivity had sought to explain US productivity losses in terms of the interaction between the internal culture of American firms, the macroeconomic policies that discourage long-term

---

[2] A report by the Office of Technology Assessment on HTS credited the US with having mounted a comprehensive R&D effort with $130 million in federal funding, but immediately cautioned against complacency. It noted that Japan had already demonstrated superior capabilities not only in high-quality materials processing and microelectronics but also in its ability to maintain long-term investment in materials research and a strong commitment from its major corporations (OTA, 1990). For a discussion of the different options in the development of HTS research, see also Simon & Smith (1988) and Rowell (1988).

production strategies, and national characteristics that operate as constraints on market forces (Dertouzos *et al.*, 1989). The MIT study identified a number of factors contributing to an American corporate culture oriented toward short-term, low-risk, high-return investments, rather than long-term increased production. In contrast, the Japanese government was portrayed as having played a crucial role in reducing risk and securing investments with government-coordinated strategic programs encouraging exports, rewarding investment in productivity growth, protecting domestic markets, subsidizing R&D, and discouraging conflicts between companies. The MIT study concluded that major Japanese corporations can thus afford more patience than their US competitors (Dertouzos *et al.*, 1989: 64).

One critic of the US response to HTS has characterized it as a "study of what might be done if something is ever discovered that might have a known commercial value" (Crow, 1989a: 341). The Defense Advanced Research Projects Agency (DARPA) was the most important US supporter of LTS and is presently an important supporter for HTS – the Department of Defense (DoD) finances half of all government-sponsored HTS research in the US. That exception granted, Crow sees no federal plan or national strategy worthy of the name. The federal HTS committee set up in the wake of President Reagan's initiative, dubbed the "wise men's committee", may have made some good suggestions, but lacked clout and money. It is said that too many agencies act too independently of each other and without clear objectives. The US system, with its focus on individuality, may be at a disadvantage when a group response is required (Crow, 1989b). The contrast between US industry's "wait-and-see" approach and Japan could not be sharper:

Japan is able to use the government as a guiding force in establishing national and industrial consensus on what technologies should be developed. Group efforts are directed at attaining technological and economic objectives, with science being used to support the primary objective. In the US it is every organization for itself and there are hundreds of plans as opposed to a few, and thousands of industrial, corporate and laboratory objectives as opposed to a few national objectives

(Crow, 1989a).

As a most careful and thorough analyst of US government science and technology policies, the Office of Technology Assessment (OTA)[3] did not share this jaundiced view of what is usually considered the strength of the United States' research system: its wide variety of institutional and funding arrangements. The OTA's second report on HTS presents a well-balanced assessment of progress and potential applications (OTA, 1990). Its forecast of a likely

---

[3] Regrettably, the OTA was dissolved in September 1995.

sequence for the development of HTS technologies draws conclusions from earlier experience with LTS. It suggests that preferred materials for applications are likely to be those easiest to handle and manufacture, not necessarily the "best" superconductors or those with the highest critical temperatures, and that highly reliable, conservative designs will be the key to the commercial success of products.

Regarding the selection of funding targets, OTA cautioned against those likely to be leapfrogged by well-entrenched and steadily improving conventional technologies. As the magnetic resonance imaging devices had already demonstrated for LTS, it is impossible to predict what HTS's most successful applications would be. Nonetheless, OTA drew up a timetable for the development of various applications. In the short term, it expected applications to emerge in the defense and space sectors, as well as in electronics and communications. These areas overlap to a great degree and high-performance considerations outweigh cost factors. In the mid-term, it expected medical and industrial applications. Only in the long term did it expect applications in electric power, transportation, and high-energy physics. The report also analyzed the US response, including a detailed overview of federal programs, their budgets, and the various coordinating systems set up within federal agencies and between state and federal agencies. HTS programs in other countries received a brief and not always accurate glance. The main focus was on comparing HTS research efforts by US and Japanese industry.

To this end, the OTA and Japan's International Superconductivity Technology Center (ISTEC) conducted a joint comparative survey on industrial superconductivity R&D in the two countries. In 1988, the amounts spent on HTS research and development in the United States and Japan were roughly equal, but funding had different sources. While the US government spent $130 million and the Japanese government half this sum, industry spending was close to the reverse: internal HTS funding in the US totaled about $74 million, with 440 full-time researchers, while in Japan it came to $107 million with 710 full-time researchers (OTA, 1990). In the US, almost half the federal funds had gone to the Department of Defense, while Japanese spending was all aimed at commercial, civilian applications. The OTA points out that while military and civilian requirements for HTS are largely the same at the present, this could change as the field matures: military and commercial R&D priorities are likely to diverge as HTS technology is incorporated into weapons systems. Another reason for US concern, especially in view of the scarcity of university resources, is that a large share of the remaining federal research support went to the US national laboratories, whose past record in technology transfer has been poor.

The OTA/ISTEC survey found that the most serious threat to the United

States's competitive position is that US industry's level of investment in LTS and HTS research and development is much lower than that of its foreign competitors. In the context of increasing doubts about US ability to compete in global high-tech markets and the recent loss of market shares in consumer electronics, memory chips, automobiles, and machine tools, this was alarming. That US industry's investment strategies are short term was attributed to severe economic and financial pressure – the shortage of patient capital. But the OTA sees no easy solutions. HTS merely highlights broader policy issues that could, in principle, be remedied. The OTA report concludes:

If US competitiveness continues to decline it will not be because the United States lost the superconductivity race with Japan, but because policy-makers failed to address the underlying problems with long-term, private sector investment

(OTA, 1990).

This is not the place for a detailed examination of the strategic and structural differences between the US and Japan in their competition for leadership in high-tech markets.[4] But one main structural difference surfacing again in HTS research lies in the role of the military as a sponsor for basic and applied research. In the 1960s and 1970s, government support was crucial in all countries doing LTS research. Targeted were commercial and in some cases military applications, such as wire and cable production, magnet winding technology, and electronics. In the United States, the Department of Defense began funding superconductivity in the late 1940s with the establishment of the Office of Naval Research. Later, the Air Force supported research in sensors, airborne generators, and signals processing, and the Navy in magnetic and electromagnetic infrared detectors and marine propulsion systems. Defense agencies such as DARPA also continued to support LTS electronics research, often as the only federal agencies to do so (OTA, 1990). In the 1950s, for example, DoD-funded research had led to the development of the Bardeen–Cooper–Schrieffer (BCS) theory explaining superconductivity. At the time of the report, DARPA disbursed more than 40% of the DoD's HTS funding; it was unique in focusing on HTS materials processing and the development of prototypes.

This raises the issue of the role played by defense in so-called dual-use technologies, those with both civilian and military applications. Some authors have drawn a functional parallel between many DoD programs with unconventional approaches to R&D and Japan's MITI (Ministry of International Trade and Industry). For defense-related programs, the linear model of technology development – first research, then development, design, manufac-

---

[4] For a detailed analysis see Derian (1990).

turing, prototype testing, and market analysis – has always been deemed too slow and inadequate. Since defense-related technologies are not market-driven, DARPA was able to use a parallel approach to expedite the development of devices. Now this approach was advocated for HTS in general (Robin *et al.*, 1988). Although DARPA had a good record in funding dual-use technologies, the question remained whether the dual-use approach was optimal for commercial applications. With the end of the Cold War and the collapse of the communist regimes, future economic competitiveness might replace military superiority as the essential factor in national security. Indeed, the Carnegie Commission on Science, Technology and Government recently proposed a number of ways in which US economic performance might be boosted by reorganizing the interplay between military and commerical R&D (Pool, 1991).

In the course of these and similar policy debates, the United States and a number of other countries have examined the potential of industrial consortia to fund research and to accelerate the commercial development of HTS. Again, HTS has found itself at the center of a longstanding policy discussion. In late 1988, the "wise men's" Advisory Committee on Superconductivity, operating under the White House Office for Science and Technology Policy, recommended the establishment of four to six superconductivity consortia, each including representatives of a major university, a government lab, and several private companies. In April 1989, the first consortium to develop superconducting electronic devices was announced. Its participants were MIT, Lincoln Laboratories, IBM, and AT&T. While other National Science Foundation research consortia concentrated on basic research, this one would direct its efforts to applied HTS research.

But this initiative was pre-empted by the 1988 founding of the International Superconductivity Technology Center (ISTEC) in Japan, a joint, precompetitive cooperative research institute financed by 111 full and associate members, five of them non-Japanese. ISTEC unites in one consortium all major Japanese companies involved in HTS R&D, organized under the auspices of MITI, which also contributes one-third of its funds. Under excellent conditions, with the most advanced equipment at their disposal, about 90 scientists do basic research on materials development. Together with their large and well-equipped industrial labs and due to the practice of funding in ten-year increments, the Japanese pose a formidable challenge to the rest of the world (FOM, 1989; OTA, 1990).

By comparison, US industry consortia follow a pattern developed in the 1980s as a popular technology policy tool. Most such consortia are privately sponsored; the joint industry–government Sematech is a notably successful exception (Derian, 1990). Following the announcement of the first US industrial consortium for applied HTS research in 1989, concern was voiced about this precedent. It was argued that, because the various consortia would work in similar areas, they could end up as a patchwork of small groups, each below the "critical mass" in size, wasting federal funds on duplicated efforts (OTA, 1990). Two years later, this fear of a proliferation of consortia proved unfounded. As an experimental organization form, it displayed both strengths and unexpected weaknesses. It removed the usual barriers to the free communication of ideas and information among researchers. The DARPA-funded consortium encouraged complete openness in research and the sharing of intellectual property. The most important factor keeping the consortium going turned out to be the graduate students who moved from lab to lab; they became familiar with the different research cultures and began to function as messengers. But the consortium has failed to induce small companies to translate technical progress into products – which was a central aim. Since property rights are more important for small than for large companies, it has proven difficult to include small companies.

Indeed, the US Consortium for Superconducting Electronics, with its members drawn from top US research institutions, is regarded as a test case for the effectiveness of research consortia in general. If it has no effect on US competitiveness in superconductivity over the next decade, it is likely the US will abandon the idea of such consortia (Pool, 1991).

Testing government policy: the United Kingdom

The United Kingdom has closely followed and imitated US efforts to improve government-sponsored cooperation between universities and industry in developing HTS. Once again, HTS was used as a test case for the implementation of current government policy. The UK example also illustrates some pitfalls encountered when a government's desire for industrial participation is not requited by industry.

By the time HTS was discovered, the UK no longer had a coherent major effort in conventional superconductivity. In the 1970s the UK had been a leader in LTS applications. The Ministry of Defence had investigated superconductivity in the context of ship propulsion, the electrical engineering committee of the Science and Engineering Research Council (SERC) had funded LTS research, and the energy industry had been working on a superconducting generator. But when no real commercial opportunities developed, these activities were given up or scaled down in the early 1980s. The SERC Material Committee and the Department of Trade and Industry had started an initiative on powder metallurgy, but this avenue, too, never achieved the expected success and was not renewed. The SERC Material Committee also failed to declare superconducting electronics one of its special research areas. A few projects on thin-film techniques were still financed by the Ministry of Defence and some basic research projects on superconductivity were supported by SERC's Physics Committee. The few programs that were left were geographically dispersed and funded by different bodies. The superconductivity research community was thus rather fragmented, with few links between the different research areas and no apparent coherent line of research or program structure (Jansen, 1994).

As the economic recession set in, British industry and university funding were in trouble. In Spring 1987, the Department of Trade and Industry (DTI) announced its willingness to provide funds to foster industry–university cooperation as part of the LINK Scheme (which aims to bring together universities, government laboratories and companies to work jointly on commercially promising projects in areas of British scientific and technological strength). Several research groups and companies were ready to take up the offer. The UK companies, "although far from leading the field of superconductivity" at the time, included GEC, Plessey, British Oxygen, and Oxford Instruments (*New Scientist*, April 23, 1987: 20). Oxford Instruments produced body scanners for hospitals and was one of the world's most successful producers of small magnets. Also, the SERC arranged the first major meeting of British HTS researchers in the hope this would improve collaboration among various groups, avoid duplication, and optimize the use of scarce resources. But the meeting ended disappointingly for those who had expected funding. Twelve groups had applied to SERC for grants totaling $2.3 million[5] – approximately SERC's total physics budget. Industry representatives attended the meeting, but as *Nature* reported, "it was felt that the potential importance of the present work was not impressed upon them sufficiently strongly" (*Nature*, Vol. 327, May 7, 1987: 4). Industry was thus unwilling to contribute. Representatives of the Ministry of Defence were also present, but made no formal commitment regarding future policies.

Industry's attitude toward investing in superconductivity did not change substantially with the discovery of HTS. According to Sir Martin Wood, chairman of the UK National Committee for Superconductivity (NCS), "The attitude of industry is that if something is being done in Japan and the United States, then Britain will not be able to keep up, and that if those countries are not involved, then it is not worth doing ...." (*Nature*, Vol. 335, September 15, 1988: 196). By September 1988, of the $27 million in public money made available, only $17 million had been spent; under the LINK scheme, funds had

---

[5] Exchange rate: 1£ = $1.70

been designated for an advanced high-tech project composed of two or three industry partners and a collaborating university institute – a cooperative effort that never materialized. Another \$8.5 million from a joint SERC/Ministry of Defence scheme also had no takers.

Meanwhile, university researchers complained about the difficulties of keeping up with international developments without the funds to build up a comprehensive data base and about the inability to attract suitable postgraduate researchers with only short-term contracts. As early as 1987, all those involved in HTS were worried about the government's squeeze on university budgets. In 1988, the nature of long-term research seemed forgotten in the drive for efficiency in universities. The general consensus was that staffing, rather than equipment, was the major problem in the UK (*Nature*, Vol. 326, April 2, 1987: 432). Nor could British industry offer university researchers anything to compare with the work being done by Bell Labs. or IBM in the US.

The course of British HTS policy was strongly influenced by the government's overall science and technology policy in the 1980s.[6] The discovery of HTS, the belief in its commercial potential, and the interdisciplinary approach to research it required made it the ideal test case for the government's longstanding desire for national coordination of research, which seemed an idea whose time had come in 1987. SERC and the Department of Trade and Industry set up the National Committee for Superconductivity (NCS) in October 1987; half the members came from the conventional superconductivity industry. The committee was to coordinate support from these two bodies and to make recommendations to SERC about the site, organization, and resources of the proposed University Research Centres for Superconductivity. Several policy concepts could now be put into operation: selectivity, specialization, and "centres of excellence". The NCS explicitly and rather crudely stated the desire for effective collaboration:

... the purpose of the NCS fits in well with the Government's wish to streamline and focus the science and technology system. The Government's recent strategy in the public funding of science is designed to reduce needless duplication by coordination of governmental, academic and industrial research institutions and to emphasize research in precompetitive and industry-relevant areas

(Tyszkiewicz, 1988).

The statement continued that the prerequisites for funding for superconductivity research were national coordination and the improvement of inter-university, inter-industry, and university–industry links. On the industry side, precompetitive, collaborative industrial research would be funded. This new

---

[6] See also Senker (1991).

policy was modeled on the industrial research consortia in the United States (Tyszkiewicz, 1988: 29ff).

What followed was a funding strategy for a three-tier system. At the top, the Interdisciplinary Research Centre (IRC) in Cambridge would receive about 45% of SERC's total budget for superconductivity (about $3.4 million annually) and was expected to drum up matching funds from industrial and academic co-sponsors. The second tier of six independent collaborative projects involving 25 university departments would receive another 45% of the budget. The final tier of 15 geographically dispersed university departments with small projects would receive the remaining 10% (Tyszkiewicz, 1988: 45). The May 1988 appointment of the retired head of an industrial research lab as the IRC's first director was clearly a political decision. Even though he was an experienced manager, he lacked a good background in superconductivity and was replaced only a year later by a well-known Cambridge superconductivity researcher. Staff for the IRC would be drawn largely from postdoctoral associates, graduate students, and technicians. As elsewhere, the focus of work would be on electronics and other applications.

The Cambridge center was the first of 12 IRCs that were to be established under the sole control of SERC (up to 300 suggestions for further centers had been received), and it was sharply observed and sharply criticized. It was feared that the IRC's programs would be dictated by industry needs, stifling initiative and presaging ill for science. But supporters were convinced that IRC's would revitalize British research, if financing were adequate and flexible local management could be guaranteed (*Nature* , Vol. 331, February 25, 1988: 648). Tyszkiewicz, who observed the British superconductivity community during this period, argues that the coordination of resources alone is not sufficient to make the national program effective. The British effort was clearly constrained by the limited public funding that has affected Britain's entire science and technology system over the last ten years. In her opinion, the coordinated program for superconductivity and the Cambridge IRC have received more government and media attention than their small budgets warrant. She sees the government's long-term commitment to superconductivity research as precarious; with HTS applications probably a decade away, the guaranteed six-year lifetime of the IRC seems much too short. Nonetheless, and though they commented on the lack of resources, many of the British researchers Tyszkiewicz interviewed considered their programs distinctive and competitive. They saw their main competitors in HTS research as Japan, the United States, Germany, and France – in that order. They saw Germany's program as the European leader, with a budget twice that of Britain's (Tyszkiewicz, 1988: 17ff).

In Summer 1991, SERC faced a critical, independent review of its first two IRCs.[7] Public knowledge now confirms Tyszkiewicz's assessment. In both centers, management and organizational troubles severely limited research productivity in the first year. The centers' management structure was distinctly "top down", supposedly to please Prime Minister Margaret Thatcher, who was eager for British participation in the international HTS competition. But with the UK economy in a recession, the appointment of an experienced, retired industry manager as director of the Cambridge IRC was not enough to attract the planned 50% industry investment by the end of six years; after two years, industry contributed only 30%. Half of this industry support had been in kind, including, for example, a 15-year-old microprobe analyzer beset with the problems to be expected from equipment that old, and the $5 million molecular beam epitaxy (MBE) offered by GEC, which the company no longer used, but which required expensive cleaning to remove cadmium and mercury contaminants (Anderson, 1991a: 270–1).

Thus Britain's intended counterpart to the US industrial research consortia does not appear to be a successful innovation. Nor does it compare favorably with German research conditions (Jansen, 1994). Industrial participation ordered from above is no more likely to succeed than when imported from abroad.

The German program to promote HTS[8]

By the time of the discovery of HTS, the German Federal Ministry for Science and Technology (*Bundesministerium für Wissenschaft und Technologie*, BMFT) could look back on two decades of funding research on the properties and behavior of intermetallic superconductors and their potential applications. Of the four major programs funded with a total of $6.5 million[9] a year, three were oriented toward applications (in medical equipment, superconducting power generators, and what were called physical technologies); the fourth, a much smaller program endowed with less than 10% of the budget, dealt with more fundamental questions in superconductivity.

The BMFT had revised its policy by the mid-1980s. Now it was judged essential to bring universities and industry together in joint projects, termed *Verbundforschung* (collaborative research). The policy aims behind this ranged from facilitating more ambitious programs otherwise too large for individual

---

[7] The Review of the UK National Superconductivity Programme conducted by the National Committee for Superconductivity was published by SERC in January 1991, but it was impossible to obtain a copy of the independent review.

[8] For the German case see also Jansen (1991, 1995).

[9] Exchange rate: $1 = 1.7 DM.

firms, institutes, or university departments, through ensuring the necessary knowledge transfer between universities and industry, to pooling financial resources so that public and private sectors could collectively choose future research options. In conventional superconductors, however, joint undertakings played only a marginal role. In 1986, 75% of BMFT funds went to industry and 25% to universities.

In addition to the BMFT, several other funding bodies provided money for superconductivity research in the universities. The Volkswagen Foundation, set up by the German federal government and the *Land* (federal state) of Lower Saxony, invested more than $0.3 million in microstructure technology and in a series of conferences on superconductivity electronics. The German Research Council (*Deutsche Forschungsgemeinschaft*, DFG) spent $2.1 million on two special research programs (*Sonderforschungsbereiche*). These are large, often interdisciplinary research projects bringing together theorists and experimentalists from several universities. The grants run for three years and are renewable for up to a total of 15 years, subject to evaluation. In addition, work on superconductivity was firmly installed at the "big science" labs, particularly the *Kernforschungszentrum* in Karlsruhe (KFK), where it focused on basic materials, the micromechanisms of superconductivity, and high-field magnet technology within the European fusion project.

In 1984, a new four-year superconductivity program had been started, coordinated by a project management agency (*Projektträger*), the VDI[10] *Technologiezentrum* in Dusseldorf. Such intermediate bodies are charged by the Ministry with pre-selection, guidance, and the subtle management and control of projects. At the end of the four-year period, discussion would ensue on whether to continue research in this direction. Under these auspices, invitations were sent out in late 1986 for an expert meeting in January 1987. The meeting would discuss the future of conventional superconductivity in Germany. So far, there were few signs of the excitement that would seize the entire German scientific community only a few weeks later. None of the German scientists had reproduced the Zurich results, and no one knew what further developments would be.

It is important to note that the structural framework of the German R&D system and its science policy profile differs greatly from that of smaller countries, Switzerland, Austria, and the Netherlands. In German R&D funding, not only the federal government is important, but also the *Länder*. To illustrate: the relation of federal to *Länder* level funding is 1.7 to 1 (Campbell, 1993: 1, Felderer & Campbell, 1995). Thus, in Germany, a scientific institution's

---

[10] VDI = *Verein Deutscher Ingenieure* (Association of German Engineers).

situation depends greatly on its exact geographic location and the general science policy of its *Land*. Further, Germany has a differentiated, well-established, and partly publicly funded network of non-university research institutions that have come to play an R&D role as important as the universities (Irvine *et al.*, 1990: 50–78).

By February 1987, the German situation began to change dramatically when physicists in Karlsruhe called a meeting of German researchers and ministry representatives to discuss the HTS news and develop a common research strategy. The meeting coincided with a highly symbolic event: the first German reproduction of the oxide superconductor, at Karlsruhe. All over the world, the Zurich results were being reproduced and new compounds with similar properties discovered. While there were still no concrete demands for immediate funding of bigger national programs, it was nevertheless clear to many of the German groups that they wished to take part in the international research effort. Only the university representatives, suffering from chronically low institutional budgets, stressed that they needed money to start. The BMFT immediately responded by allocating about $175 thousand to 30 universities to allow conference attendance, purchase chemicals, and cover increased communications expenses.

The concerted research efforts taking shape in many industrialized countries, particularly the United States and Japan, with the intent of staking claims in future markets for high-tech products made it obvious that a special kind of coordinated program would have to be set up in Germany as well. To underscore the urgency of the need for funds, researchers cited the example of Siemens; this German company had become one of the world's major producers of nuclear magnetic resonance imaging devices, an application and economic development unforeseen when Germany began conventional super-conductivity research in the 1970s.

Indeed, industry – in particular, Siemens and Hoechst – became a major actor in setting up Germany's HTS effort. Siemens, interested in the energy and microelectronics markets, could already look back on a long tradition in conventional superconductivity research and was already a market leader in LTS devices. Hoechst, by contrast, was a newcomer from the chemical industry. Prior to 1987, the company had never worked on superconductivity, but it hoped that its expertise in manufacturing ceramics would compensate for this lack of LTS experience.

Siemens conducted superconductivity research at its Central Research and Development Unit (*Zentrale Forschung und Entwicklung*), where two units in the Department for Technical Physics dealt with conventional superconductivity. No less than 75 researchers were involved in this field. In the 1970s, they had examined NbTi and Nb$_3$Sn, and Siemens' subsidiary Vacuumschmelze had used these materials to become Europe's largest producer of superconducting wire. The applied superconductivity unit, together with a group in Karlsruhe, had been developing high-field magnets for use in scientific research and in building a superconducting generator and a resonator. In the 1980s, Siemens' efforts focused on developing magnets for nuclear magnetic resonance imaging devices, which turned out to be the most important application of superconductivity. This research was funded in the framework of a BMFT program on medical imaging technology. As in many other superconductivity research projects, basic research was sacrificed in the rush to find lucrative applications. Having realized this omission, by the end of the 1980s, Siemens initiated a complete restructuring, putting more weight on in-house basic research in technologically important sectors again.

As a well-established firm in the area, Siemens had longstanding collaborative links to university researchers and Germany's "big science" labs during the precompetitive phase of the innovation process. For example, it had a contract on superconductivity technology transfer with one of the Karlsruhe institutes. Siemens "knew the ropes" of the superconductivity network. Yet, despite its advantageous position, the company did not join the race until March 1987. Premature reports of HTS had often shaken the field of superconductivity before. However, once the reports were confirmed, it was considered inevitable that Siemens would join.

A central problem in applicability also dampened researchers' excitement; the high degree of anisotropy of the ceramic materials kept current-carrying capacity low. Much basic research would be needed to overcome this problem, and marketable applications were probably at least ten years away. But, as one Siemens group leader explained to us, the field was so new and so fundamentally different from what had been done so far that it was crucial to participate from the beginning, or risk no longer being a player when the development of applications began. The lucrative microelectronics sector, specifically high-density computer cables, held the greatest promise.

When HTS research began in April 1987, Siemens chose 35 collaborators with experience in conventional superconductors as well as applied materials research. Siemens' two subsidiaries, Vacuumschmelze and Interatom, joined the effort in order to protect the "old" superconductivity markets in magnet and medical technology as well as in hopes of reviving the energy technology market. Being still in a precompetitive phase, Siemens' first step was to intensify its contacts with universities.

The second German company playing an important role in the formation process of HTS as a research field, was Hoechst. Informed by the February

1987 meeting on superconductivity in Karlsruhe, Hoechst's senior research management, without formal approval from the company board, had decided to launch an immediate HTS program that same month. This would allow them to form a better judgment of the field's potential and the prospects for long-term investment. Hoechst Celanese, in the US, also committed itself to a research program on the new materials.[11] Hoechst had not been involved in conventional superconductivity because it had no expertise in metals. The new superconductors were ceramic – right up Hoechst's "alley". As one of our interview partners noted, Hoechst had "know-how in ceramics, [knew] how to produce, press, and sinter powders ... [had] know-how in the physics of superconductivity, ... sputtering know-how, ... [and] in making layers". Hoechst also had the commercial strength to market potential products. The aim was clear: to secure a good starting position for future markets.

Hoechst's high regard for basic research is typical of Germany's chemical industry. Industrial researchers have excellent contacts and a long tradition of cooperation with universities. As a newcomer to HTS, Hoechst had to act quickly, without building up an in-house specialist team first. "Some people were told to stop what they were doing immediately. This was a completely new procedure at Hoechst." But shorter production cycles for high-tech products mean less time to amortize R&D costs, so it was impossible for one lab in one enterprise to cover all aspects of HTS research, and intense cooperation with outside researchers became essential. Because HTS research was considered a risky venture, several separate contacts with external groups were set up to enable several research directions to be followed in parallel. Cooperation with competing industries seemed out of the question at this stage (an attitude that would soon change), but within two months, Hoechst formed a project group consisting of six scientists and signed more than ten exclusive cooperation contracts with university groups.

The creation of this kind of network strongly influenced Germany's entire superconductivity community. It followed the model of coordinated industry–university projects funded by the BMFT, but with an important difference. Hoechst was a single private company at the center of the cooperation cluster linking all the other groups. This network recruited brilliant young researchers and, by the end of 1987, the team had 14 members, one-half chemists and one-half physicists. The outside cooperation partners were members of various BMFT programs, the Volkswagen Foundation program, or DFG special research programs. They covered aspects of materials science, conventional

---

[11] In the June 1991 issue of *Physics Today*, Hoechst Celanese offered its services in producing tailor-made Bi- and Y-superconductors.

superconductivity, and production technologies. Hoechst thus gained access to a large variety of equipment and know-how. Moreover, HTS research was not confined to one R&D unit, but was organized across structural borders without problems.

In addition to the universities (the most enthusiastic proponents) and industry, non-university research labs also played a very active role from the beginning. Non-university labs' HTS research efforts varied widely in size, position in the research domain, and orientation toward basic or applied research. While it is difficult to make a more general judgement, two points should be noted. Non-university labs held an important position in the German science system in negotiating the national programs, even though they could not or did not need to profit from these financial resources themselves. And they made numerous and crucial contributions to research itself, reflecting their size and importance in Germany. Most plunged into HTS before the joint university–industry projects; this reflects the advantage of financial independence and stable organizational structures. Our detailed investigations concentrated on the two "big science" laboratories, Karlsruhe and Julich. Among the Max Planck Institutes, we chose the one in Stuttgart.

When HTS was discovered, Karlsruhe was the center of German superconductivity research. Two of its institutes were strongly involved: the Institute for Technical Physics (ITP) was applications-oriented and worked on high-field magnets for nuclear fusion experiments, while the Institute for Nuclear and Solid State Physics (INFP) had become the center for superconductivity basic research when many other labs reduced their efforts in the field.

Embedded in the same institutional context and with similar starting positions, the two Karlsruhe institutes nevertheless responded differently to the discovery. The basic research group joined the international effort almost immediately, the applications-oriented team more reluctantly. Applications were too far off, and the researchers were already committed to a number of well-defined projects such as developing magnets for particle accelerators and fusion reactors. These could not be abandoned on short notice in favor of attractive new projects, since such long-term research contracts are essential for the survival of the institute. A "big science" lab functions much more like an industrial enterprise than like a university research unit. In a sense, researchers are responsible for developing a product: they hold a kind of share in the science–technology market that guarantees the financial health of the institution. The INFP researchers began work as early as January 1987 and were the first in Germany to duplicate the Zurich experiments, reaching 42 K in the LaSrCuO system.

The second "big science" laboratory in Germany to engage in HTS research, Julich, was also experienced in conventional superconductivity, though it had abandoned the area in the late 1970s. When HTS was discovered, Julich was in a phase of organizational restructuring and reorientation. The existing Institute for Solid State Research (IFF) was devoted to basic research; a new Institute for Sheet and Ion Technology (*Schicht-und Ionentechnik*, ISI) was set up specifically for applied research. Both became involved in HTS research. With suitable equipment and the expertise of superconductor specialists who had remained at the institute even after their specialty had been phased out, the IFF was able to join the HTS effort as early as 1987; the new institute would join only later.

As at Karlsruhe, the aim here was not to create one strongly coordinated project, but to tailor HTS research so that it could be pursued within existing programs, particularly the one on basic research for information technologies, which had three main research directions. Potential applications of HTS were studied under the heading "solid state research for information technology"; basic HTS research was conducted in "cooperative phenomena in condensed matter"; and materials preparation fell to "alloys, phases, microstructures, and diffusion". Though no coordinated effort had been planned, it arose as a result of the topic anyway. Four of the IFF's ten units began making and testing thin films and superconducting layers. Another group working on ion implantation and analysis would later move to the ISI to tackle the production of thin layers using laser and ion implantation methods. Its long-term interest was the development of hybrid superconductor/semiconductor devices. Finally, a team of analytical chemists added an interdisciplinary balance that had been lacking in the Karlsruhe team. Cooperation was strongest in the "solid state research for information technology" group.

In this early phase, the Max Planck Institutes were in quite a different position. They are not funded by the BMFT, but by the Max Planck Society, which negotiates its budget at the federal and *Länder* levels each year. But allocations are made far in advance, making it difficult to respond rapidly to unexpected events such as the discovery of HTS. The Max Planck Institutes felt no formal external pressure to collaborate with other groups or to fulfill any preconditions to join the national effort. Some groups at the Max Planck Institute for Solid State Research in Stuttgart, for example, started early, but went it alone, using their expertise to establish independent niches in the field. Specialized in semiconductor physics, this lab was a newcomer to superconductivity. In January 1987, theorists in the lab spurred the chemists to react, followed by experimental physicists in spectroscopy. An HTS project group with some twenty to thirty researchers was set up a month later; it had broad know-how in materials, excellent international connections, the required qualified manpower, and the equipment to begin high-quality, internationally competitive research immediately. Its interest in the research effort was in gaining a fundamental understanding of the phenomena.

On the national level, the BMFT – advised largely by members of the old superconductivity community – announced in June 1987 its readiness to invest in this new research field. Basic research on new materials and production techniques would be funded for a period of three to five years, information exchange would be intensified and accelerated, and the coordination of the interdisciplinary groups involved would be improved. A later phase would concentrate on utilizing the expertise gained to produce "magnets, cables, energy storage units, and generators".[12] The national science budget for 1988 had already been allocated, so new funds were needed. The proposal to increase superconductivity budget was brought before the *Bundestag*, the lower house of Germany's parliament, in Summer 1987, just when competition on the level of international politics was at its peak (with US President Reagan's announcement of the superconductivity initiative). After consulting with leading superconductivity experts, including industry representatives, the Ministry and the program-managing agency acquainted with conventional superconductivity drew up a plan to fund HTS research. The budget for applications-oriented basic research would be increased by $7 million. The VDI *Technologiezentrum* spokeswoman responsible for the HTS program said that superconductivity was clearly regarded as a long-term research topic. In the first phase, the main concern was to rapidly build up interdisciplinary research teams that could form the basis for broader basic research, as well as for applied research in cooperation with industrial enterprises. HTS would be funded within the BMFT program "Physical Technologies", money also being transferred from LTS to HTS research.

The BMFT's response was non-routine in two ways. First, it decided to fund local interdisciplinary cooperations, including researchers based in different university institutes. The intent was to create new cooperative links that would continue well beyond HTS research. Second, it decided to fund only university institutes in the first phase. German industry was invited to participate in the new program, but would have to finance part of its university partners' research. Normally, the BMFT gives priority to industry or joint industry–university ventures, rather than to individual university departments. Why did the Ministry act this way, and why did industry accept it, at least initially?

---

[12] BMFT press release, June 1987.

First, everyone involved saw that the money available in the first year would not be enough to adequately fund both industry and universities. Second, there was general consensus that basic research was at the center of the program's first phase and that applications were still remote. Third, the relevant expertise was mainly available in the universities. Industry would need some time to establish networks to gain access to this know-how. Fourth, university priority in financing in the first years gave industry confidence that the BMFT would embark on a long-term project, in whose framework industry would later also find good funding opportunities. Fifth, it was important to reach agreement on funding without losing time. Finally, a strong, interdisciplinary university research program would provide industry with a pool of know-how at a later stage – at no risk or investment.

But industry was not unanimous in greeting these decisions. Hoechst, for example, had hoped the BMFT would fund HTS in the physical technology program, rather than as materials research. Second, the investment in this first phase would not add to the industrial projects which were postponed to a later stage. Hoechst had already invested heavily in the research, and feared that the definition of research aims in the BMFT physical technology program would not fit the research objectives of a chemical company. Hoechst usually did research in the framework of the materials research program and was familiar with the "customs" adopted in this research sub-community. The German HTS program focused on the utilization of the new phenomenon, while Hoechst was more interested in basic research and development of the materials. Hoechst saw the BMFT program as too oriented towards technological and economic considerations about magnets, thin films, and other potential devices, rather than on the materials themselves.

Hoechst continued to try to convince the Ministry of the importance of its basic research program, hoping that funding could be secured through its university partners, if not directly. And, although it was agreed that the first phase of research would exclude funding for industry, this policy changed quickly. The Ministry, whose policy was to foster knowledge transfer from universities to industry, found it difficult to withstand industry's increasing pressure. Siemens and Hoechst argued that other countries' governments were meanwhile generously funding industry research, and that the exclusion of German industry would disadvantage it in the global competition for future applications.

As a result, in September 1987, the BMFT announced a second call for proposals, this time foreseeing funds for industries collaborating with universities. But no money was available in the 1988 budget and the BMFT was unwilling to redistribute money allocated to the universities, so industry received only "letters of intent", with promises of funding in later years if the budget permitted. Industry had to wait until 1989 to receive full funding for its HTS research.

Meanwhile, the field's global situation had changed. In the light of the competition between the US and Japan for leadership in the race for new materials and potential applications, in the summer of 1987, Hoechst and Siemens began discussing a possible cooperation. Both believed a joint German industry effort would strengthen their position in the world market. Although interest areas overlapped to some degree, both regarded HTS research as being still precompetitive and neither expected this to change until applications appeared on the horizon. Hoechst was attracted to such a cooperation because it meant allying with a group already well-established in the conventional superconductivity network and because it would increase the likelihood of receiving public funding for other research in the future. Meanwhile, a third potential partner, Daimler Benz, also announced its interest, bringing with it a number of links with other specialties, including superconducting cables, high-speed trains, etc.

While negotiating among themselves, the three industrial partners also applied separately for BMFT funding. Since their interests overlapped, they were asked for guarantees of coordination between their chemical and materials research and physical engineering sectors. In spring 1989, after a full year of negotiations, Siemens and Hoechst signed a first cooperation agreement and gave Daimler Benz/AEG an option to join later.

To ensure a smooth start for the BMFT national program and to allow informal negotiation, applicants were asked to first submit rough funding proposals to the project management agency (VDI *Technologiezentrum*). This would enable the BMFT to see who wanted to do what, where, and with what means. After a preselection, VDI coordinators visited the labs to negotiate staffing and equipment requirements before official applications were formulated. This would minimize the administrative effort involved in the application process. The response to the program was overwhelmingly positive and there were numerous good applications; by fall 1987, eleven joint university research projects had been launched.

It was assumed that a one-time investment would be insufficient, so the parliamentary Committee for Research and Technology asked the BMFT to formulate a long-term strategy to encourage superconductivity research. In December 1987, a committee of experts was set up to advise the BMFT in formulating a research and funding plan. It consisted of four industry representatives (from Siemens, AEG, Philips, and Hoechst), research managers from the big science labs, and two university researchers. A primary guideline was that research should orient itself much more toward applications. This change was a response to other countries' – particularly the US' and Japan's – heavy investment in HTS research; the German government did not want to miss the opportunity for German firms to join a high-tech race.

How did the other funding agencies, such as the Volkswagen Foundation and the DFG, respond to the BMFT's initial strong commitment to basic HTS research? The Volkswagen Foundation's financial means are modest compared to the BMFT or even the DFG. Its policy is to fund exceptional projects or create incentives – always for a limited period only. The Volkswagen Foundation had financed an ongoing project in cryoelectronics that was nearing its end. The Ministry's initiative to massively fund HTS research was seen as a good opportunity for the cryoelectronics researchers to continue, so that the Volkswagen Foundation could withdraw from this research field. But this decision was overturned in 1988, and a new cryoelectronics project was launched, complementing the BMFT program.

The DFG is normally the obvious address for researchers seeking funding for basic research. But negotiations between the Ministry and the scientific community had already begun, so the DFG took a waiting stance. The Ministry wanted a well-coordinated, focused HTS program, while the DFG's "bottom up" policy is to avoid major interference in the structure and content of research programs, except in the special research programs (*Sonderforschungsbereiche*). It did set up two of the latter complementary to the BMFT program, financing superconductivity research for a three-year period with the possibility of extension. Particular attention was paid to fostering collaboration between theorists, who had little place in previous BMFT-funded projects, and experimentalists, with support also given to a chemistry-oriented program.

From 1988 to 1990, the BMFT provided no less than $61 million for research on HTS and related work in universities, research institutes, and industry. Industry spent an additional $24 million in the same period. In 1990, for example, on top of the money going routinely to the two "big science" labs at Julich and Karlsruhe, most of the additional funds went to research on superconducting materials, directed toward the basic understanding of the phenomenon, cryoelectronics, and high-frequency applications. The policy of massively funding HTS research continued, and in April 1989 the BMFT announced it would devote some $26.5 million annually to HTS research from 1990 to 1995. Plans allowed for small annual funding increases, for a total of about $200 million over the six-year period. These funds would finance research with a complex, interdisciplinary approach that required collaboration between industry, non-university research laboratories, and universities. Work would continue to focus on joint research projects, with special emphasis on university–industry cooperation. Industry and research establishments doing R&D on potential HTS applications would both receive funding.[13]

4.2 The smaller European nations: the Netherlands, Switzerland and Austria

The Netherlands

In 1987, the status of scientific research in the Netherlands was not fundamentally different than in other countries its size in Europe. Several research groups, especially at the universities, continued the strong Dutch tradition in low-temperature physics. Research of high quality on flux pinning, for example, was pursued at the Kamerlingh Onnes Laboratory in Leiden; a group in Delft was working on Josephson junctions; and the University of Twente maintained a strong tradition in work on applied LTS, building superconducting magnets for Dutch and foreign companies, including MRI magnets for the Medical Systems Division of Philips, as well as rectifiers and wires. In 1970, the government-sponsored Dutch Energy Center (ECN) in Petten began work on applied superconductivity, producing niobium–tin wire (Nb_3Sn), then the newest material in magnet technology. The ECN held several international patents and eagerly sought industrial partners. Philips, sometimes called "the only company in the Netherlands", had been engaged in a broad spectrum of LTS research, ranging from Josephson junctions and niobium–tin wire to superconducting generators and work on the superfluid properties of helium, spurred by the company's interest in dilution refrigeration. But Philips was not in the business of fabricating magnets, and as the prospects for LTS technology declined in the mid-1970s, financial considerations dictated a halt to in-house research on conventional superconductivity. At the end of the 1970s, theoretical physicists and other research groups at Philips shifted their attention to the rising field of semiconductors.

In the Netherlands, as elsewhere, basic research in superconductivity survived at the universities. But due to the peculiar structure of Dutch industry with Philips' special position, links between basic research and industrial applications were weaker than in Germany or Switzerland (which had benefited from CERN's (the European Centre for High-Energy Physics, in Geneva) orders for magnets). The ECN had attempted to collaborate with industry, but experienced great difficulties. Philips was a notable exception in

---

[13] BMFT. Supraleitung – Förderkonzept und Zwischenbilanz. Bonn, April 1991.

its willingness to cooperate with the universities, and thus became the focus of attention when HTS was discovered.

University groups began working with the new materials, but considered themselves poorly equipped. The expensive apparatus needed for ultra-high vacuum film deposition, Auger spectroscopy, or X-ray photoemission spectroscopy, for example, was not adequately accessible at the time. The scientific community was convinced that now was the time to demonstrate the importance of solid state physics and materials science, to seize the opportunity to catch up. One of our interview partners said, "People discovered that the game was more complicated than baking little samples. You could not make it alone. If you wanted to receive money, you had to stick together."

The same director who had once ended Philips' LTS research called an opening meeting in March 1987 at Philips National Laboratory in Eindhoven. It attracted the solid state physicists and physical chemists from within Philips and about sixty university scientists. It was apparent that the Philips researchers had less know-how and knowledge than their university colleagues. So the company wanted to obtain rapid access to this knowledge. After the meeting, a formal decision was made to enter the new field; this decision turned out to be essential for the ultimate success in setting up a national research program. By the end of 1987, Dutch researchers had secured $5 million in funding for a national research program for an initial period of two years. Policy-makers responsible for university research realized that no headway would be made at the national level unless Philips and the rest of Dutch industry were included in the program.

The next decisive initiative was taken by FOM (*Stichting voor Fundamenteel Onderzoek der Materie*), the research council responsible for basic research in physics. After a joint Philips/FOM conference in June, the director of FOM assembled the leading research groups from his organization, from SON (*Schikundig Onderzoek Nederland*, the sister organization for basic research in chemistry), and representatives of relevant industrial groups. Two groups were set up to work on thin films and bulk materials, coordinated by a steering group also charged with formulating a research funding proposal to be submitted to the government.

The "industrial connection" was essential to qualify for funding from the Ministry of Economic Affairs (as it normally funds only industrial research), and the industrial emphasis was reflected in the composition of the steering group. The anticipation of obtaining funds made it possible to impose certain conditions on the researchers. The group was chaired by a scientific advisor and HTS coordinator from Philips Research. Other members came from industry or relevant public bodies with links to industry. The universities were represented by FOM and SON. The funding proposal FOM submitted to the Ministry bore signatures of representatives from nine Dutch organizations[14] – everyone who counted had been included.

Like other countries, the Netherlands sets up national programs for sectors it considers to be of strategic technological importance: these include materials technology, biotechnology, and information technology. The materials technology program, for example, is funded with $33 million a year, two-thirds of which goes to industry. Another, smaller fund is reserved for innovative research programs (IOPs), which goes mainly to university research in the same area. Declaring the new HTS initiative an IOP would have involved a lengthy formal procedure; instead, the industry division of the Ministry of Economic Affairs was persuaded to commit $4 million to the program. The comparatively poor Ministry of Science and Education was also persuaded to make a reluctant contribution of another $1 million. By the end of 1987, almost a year after the discovery of HTS, a Dutch national program was underway. For two years, it would help universities upgrade their equipment. After that, HTS research funds would be available through normal channels. Because a large share of the funds came from the Ministry of Economic Affairs, the program's industrial orientation was not only strategically important, but the essential prerequisite for its launching.

How deeply committed was industry? Wild hopes for HTS applications burgeoned in the Netherlands, as elsewhere. But it soon became clear that not all the industrial participants in the new national program were actively involved. Except for Philips, industrial interest faded quickly. KEMA was interested mainly in the power engineering applications of HTS, which clearly would not be realized for many years. NOVEM, an intermediary organization established to develop national energy and environmental policy advice for the government, was attracted by the potential benefits for energy conservation and efficiency. NOVEM does not conduct its own research, but commissioned ECN to produce a feasibility study of possible HTS applications in the energy sector. The ECN study's findings were not altogether positive; it advised "keeping an eye" on future developments, but not increasing direct involvement. The ECN's participation in the national program was crucial in justifying the funds from the Ministry of Economic Affairs, but remained small-scale. The ECN saw two options for its involvement: the new program could be organized within its 20-year-old conventional superconductivity program, or it could be incorporated in its ceramic program. It chose the latter. Thus, while nominal industry involvement helped provide political legitima-

---

[14] AKZO, ECN, FOM, KEMA, PEO (and its successor NOVEM), Philips, Shell, SON, and TNO (Organisation for Applied Scientific Research).

tion for the national program, actual industry efforts were minimal, with the exception of Philips'. This accurately reflected Dutch industry's research capacity.

The steering group performed its political function well. But the divergent outlooks, interests, constraints, and agendas of university and industrial researchers soon became apparent. Industry wanted rapid, tangible results like superconducting coils, a high-current magnet, or spin-off effects in powder technology. From this perspective, the national research program had poorly defined objectives. One critic said it was a discipline-pushed, rather than a target-oriented program. The dilemma, of course, was that to "sell" the program, it should be target-oriented, while the question arose whether it was not too early for such specification. The second phase was target-oriented – though it is debatable whether these were real targets in the industrial sense. Philips occupied the middle ground, pushing for fundamental research toward applications in the field of microelectronics. Philips wanted to boost its own research program by tapping university knowledge and by establishing its own university network.

University scientists, on the other hand, are motivated by curiosity. Their steering group spokesman argued for immediate funding for laboratory facilities, equipment, and staff to enable basic research to get off the ground. Thus, a newly appointed professor and his HTS research group were able to build up a thin-film lab, where ultra-thin films and arrayed Josephson junctions are now fabricated. In his opinion, the Dutch advantage is the expertise of its fundamental research groups, though many were not conscious of any need to develop technology on the border between basic and applied research. Our Dutch interview partners with experience in the Netherlands and abroad said a key point was to accept that technology and applied research had an important role to play in science, and that it was essential to develop the skills needed to bridge the gap between fundamental and industrial research. Another Dutch advantage was its long tradition in solid state chemistry and the good record of cooperation between solid state chemists and physicists. But the disparate disciplines also needed mediation: should one train a good superconducting physicist in materials science or rather a ceramics expert in physics?

Despite these problems, and despite notorious difficulties in setting up collaboration among university research groups, it got off to a good start. FOM was accustomed to managing programs of this kind, which requires diplomacy in fostering good relations between individual researchers and research groups. Though Philips initially lacked much of the necessary expertise, the company soon developed an effective infrastructure for producing and measuring superconducting samples. High-quality thin films produced by a young Philips researcher were generously donated to university groups to help them begin measuring. FOM also intended to strengthen the cooperation by funding a university institute to set up a SQUID magnetometer for common use, though the long delivery time meant it could not be tested for months to come. In the Netherlands, cooperation generally either worked well or not at all.

Dutch university researchers were often suspicious or even contemptuous of German efforts to set up formal coordinated networks. FOM's instrumental role in helping Philips organize its own university network was openly acknowledged. Philips' generous distribution of high-quality samples made its lab central and catalytic for cooperation among university groups. But the partners took the realistic view that the cooperation was based in reciprocity; it would continue only as long as both benefited. It could not last forever if industry was involved; if commercial opportunity arose for one or the other partner, the form of cooperation would have to change.

Recruiting optimally qualified researchers is essential in building a national research program, but in the case of HTS there were no clear criteria to judge qualifications. "Sometimes things go slowly because the only way to get on a train is if you have a ticket. If you don't, it takes time to buy one," we were told by one interviewee. Not all university groups were interested or prepared to get on board. Some ceramicists had recently received a substantial grant committing them to a program launched before the discovery of HTS. Other groups happened to have a ticket and seized the opportunity. Among them were research groups who had continued to work on oxides when their colleagues had switched to metals and thin films. Those involved in work on the Jahn–Teller effect were also in a good position to shift to HTS. One leading researcher who had established a group engaged in electron spectroscopy and surface physics thrived in the national program. The group working on flux pinning and the theory of critical currents could apply their techniques and ideas to HTS. Hydrogen moving rapidly in a solid is mathematically analogous to the movement of flux quanta in a superconductor, so work on hydrogen in metals was also a ticket to climb aboard. Mastery of crystal-growing techniques was also easily applied to ceramic materials. But one group long involved in conventional superconductors and Josephson junctions was sidetracked by discovering a new effect in semiconductors.

Such accounts reveal the flexibility university groups enjoy in selecting research topics. This helps explain FOM's interest in bringing them together with Philips. The prerequisites for a decision to shift to a new field are knowledge, technical expertise, and the right equipment. None of the Dutch university groups was engaged exclusively in HTS research, except one professor who had just taken a new position. Researchers' ratings of their own share of attention to HTS ranged from 20% to 60%. The speed of response to HTS varied enormously. Those who had entered the national program would also enter the more focused second phase. Those left behind were either not interested, too slow in getting their grant applications in on time, or had proposals not considered good enough.

Before the steering group formulated the second phase of the Dutch HTS program for another round of funding, the Ministry of Economic Affairs sent a delegation to visit Japan in spring 1989. It sought information on Japanese companies' and funding agencies' choices and expectations in funding, cooperation, and organizational strategies in the HTS field. The delegation wanted to learn about current research and technology trends, and especially about the Japanese assessment of the future applications of HTS in electronics and electrical engineering. The delegation's official recommendations did not fundamentally differ from those of other observers, but the timely first-hand look at Japanese developments was instrumental in shaping phase two of the Dutch national program, especially in regard to the longer stimulation period of phase two and the emphasis on devices. The visit also reassured Dutch HTS researchers that they were progressing comparatively well in a number of areas, particularly in studies of the electronic structure of superconductors, crystal preparation, stoichiometric techniques of new compounds, and flux-pinning phenomena. The delegation found that Japan's basic and applied research was not scientifically superior to that performed in the United States or Europe; the strength of the Japanese lay in the social organization of their industrial and scientific efforts (FOM, 1989).

The Japanese research system fascinates scientists in the Netherlands and elsewhere. Several of our interviewees had visited Japan before or maintained working relations with Japanese firms and colleagues. The fascination is rooted partly in the difference in approaches. The Japanese are seen as taking a long-term approach to financing and goal-setting, leaving all avenues open to accommodate unforeseeable developments. Accordingly, their HTS programs are not as focused on specific targets, but cover a broad range of desiderata. The Japanese had taken up techniques such as metal organic chemical vapor deposition (MO CVD), not used to a comparable extent in Europe, and were now pushing ahead with rapid advances. "They do not feel the need to make decisions, which of course are always biased by what you know and what you are doing at the moment," was one comment we heard. Japanese and European approaches also differ in that the Japanese do not as easily succumb to the fascination of purely academic or conceptual problems, but always have technological applications in the back of their minds. Japanese persistence and willingness to do routine work also enhance the country's competitiveness.

In Japan, relations between universities and industry are quite weak, largely due to the weak financial support for university labs But inter-industry relations are excellent within a given company, and since most Japanese companies are large, they cover a broad spectrum of products and R&D. Links between companies are restricted to conference meetings. At ISTEC, the Japanese national-level HTS research consortium lab, each participating company is represented by one or two researchers. Tanaka's influence has helped establish good contact between them and the universities within ISTEC. The Japanese example in HTS underscores what Müller has also repeatedly said: Western scientists should question the widespread assumption that a phenomenon must first be understood before it can be put to industrial use. "The Japanese may not understand the phenomena [either], but they are able to sell what they make," said one of our interviewees. New developments in high electron mobility transistors (HEMTs) were a case in point, as Philips knew only too well.

Dramatic events in the Netherlands soon confirmed this assessment of factors enhancing industrial competitiveness on industrial markets. In September 1990, just two and a half years after the national research program was launched, Philips announced cuts in its research program, including the termination of HTS research, for financial and marketing reasons. The company also completely abandoned the fields of very advanced circuit technology and random access memories. By now it was obvious that the short-term prospects for achieving room-temperature superconductivity were poor, and that HTS materials would not be easily integrated with silicon, a precondition for advances in "cold electronics". Resources for basic exploratory research – i.e., research not likely to result in products within five to ten years – were limited and tough choices inevitable. HTS was not accorded high priority, and a plan for a group of some fifteen researchers to continue work on cold electronics applications, such as fast switches, was scrapped in favor of a small, low-key effort. Five to ten scientists would continue work on HTS, LTS, and phenomena related to "sandwiching" layers of superconductors and semiconductors.

These decisions were made in July 1990, after the steering group had submitted its funding proposal for the second phase of the national program (NOP-HTSII), again to the Ministry of Economic Affairs and the Ministry of Science and Education. The program was to span the five years from 1990 to 1994. The proposal requested $6 million from the two ministries and another

$8.4 million from the research councils. Phase two had more sharply focused research goals and an improved organizational structure. It would follow a threefold strategy: superconducting electronic devices; flux pinning and critical currents; and new HTS materials. The participating research groups consisted of a core of groups engaged primarily in research and a peripheral ring of groups with support functions. The participation of university groups was to be coordinated through FOM and SON, who would also contribute matching funds. Detailed descriptions of phase two's research aims and their coordination reveal the program's high level of internal integration. Phase two was also consolidated; of the initial twelve sites for HTS research – some of them meagerly staffed – only six universities remained in the inner ring, three of them working on more than one spearhead.

Philips' research cuts seemed to jeopardize phase two of the national program. But the momentum from phase one was sufficient to keep going, even without industrial participation. At the end of 1990, the Ministry of Economic Affairs granted a modest $3 million. With additional funds from FOM and SON, the program proceeded at a reduced level of activity and much further removed from industry than envisioned. The gap left by Philips' withdrawal led to greater involvement in European Union programs, partly to gain funding. In 1991, Philips, Thomson-CSF, and several university groups took part in the EU's ESPRIT program and in a SCIENCE program on theoretical modeling of electronic correlations. Twelve university groups, three of them Dutch, were involved in another SCIENCE program on flux pinning and critical currents. Other Dutch participants including ECN took part in BRITE-EURAM projects on ceramic superconducting filaments and ribbons. EU programs were perceived as a source of funding, but their attractiveness was marred by the amount of bureaucratic paperwork they entail. Dutch researchers were popular with foreign colleagues in many other forms of collaboration as well.[15]

Switzerland

When Müller and Bednorz made their discovery, the Swiss research system could look back on several decades of experience in both basic and applied research on conventional superconductivity. For many years, the latter had been part of the national scientific program funded by the Swiss National Science Foundation (SNF), which had been set up in 1952 to promote

[15] ESPRIT = European Strategic Program for Research and Development in Information Technology; SCIENCE = Stimulation des Coopérations Internationales et des Echanges Nécessaires aux Chercheurs Européen; BRITE-EURAM = Basic Research in Industrial Technologies for Europe.

high-quality research. After 1973, no new findings in materials research had advanced the field, so the SNF steering body, the *Forschungsrat*, accorded only low priority to superconductivity. It was placed in the *Plafonierungsgruppe*, which meant that its funding was frozen at the 1984 level. The SNF's research plan for 1988–91, formulated in 1986, no longer explicitly mentioned superconductivity at all. The conditions of previous plans would therefore continue, with funding at about $1 million per year.[16] As elsewhere, research began to orient itself toward possible applications. Since mid-1986, Swiss researchers had taken part in a joint European Community EUREKA[17] project with Plansee (Austria), Spectrospin AG (Switzerland), and the University of Nijmegen (Netherlands) to produce superconducting wires and coils using lead–molybdenum sulfides, the so-called Chevrel phases.

Swiss industry also had longstanding traditions. One of the most important enterprises was Asea Brown Boveri (ABB), a Swedish–German–Swiss group with three research centers, at Vesterås, Heidelberg, and Baden. Low-temperature research was concentrated in Baden in a few small projects collaborating with the Paul Scherrer Institute in Zurich. There was also a well-established association (*Arbeitsgemeinschaft*) of several small Swiss firms specialized in the production of Nb_3Sn and NbTi filaments; ABB sold magnets made from these filaments all over the world. Though superconductivity was not a central activity for ABB, the company decided to invest on the expectation that in the long run the phenomenon would find a place in the huge market for electrical equipment. ABB had only recently decided to concentrate on materials processing, leaving materials research to other laboratories. So work there was re-organized around product development: equipment for power stations, power transmission and distribution, turbochargers, communications, and information technology.

When HTS was discovered, Switzerland was in an enviable position. To the long tradition and commitment in the field came the fact that Alex Müller was Swiss. In contrast to Austria, where researchers had to mount tremendous efforts to convince funding agencies, Swiss funders were eager to take the initiative. They took a first step in early 1987, when the *Schulrat*, a public advisory body responsible for Swiss universities, the two *Eidgenössische Technische Hochschulen* in Zurich and Lausanne, and non-university institutes like the Paul Scherrer Institute approached researchers. But the *Schulrat*'s interest was primarily applied research, which was definitely too early at this stage. Also, the political approval process for a program launched on this level would have been too slow for such a quickly moving field. So the decision was

---

[16] Exchange rate: $1 = 1.5 SFR.
[17] EUREKA = European Research Coordination Agency

made not to fund research but to gather information on potential applications to maintain preparedness.

A second initiative came soon, and turned into a definite program. In June 1987, shortly after news of the discovery had spread internationally, the SNF gave a clear sign of willingness to finance research in the new field. It argued that scientific breakthroughs in the field of superconducting materials science warranted a stronger temporary investment. This position was not seen as contradicting the priorities set in the previous, 1983–87 superconductivity program, since LTS and HTS were not identical.[18] Basic HTS research would be subsumed under the SNF's more flexible programs, and the first program phase would focus on basic research. This fundamental decision still left open whether funding should be provided within the framework of Section 2, which normally handles one- to three-year projects under the heading "science and engineering" (*Natur- und Ingenieurwissenschaften*), or under Section 4, which is responsible for long-term national programs. For the latter, approval would be required from the *Bundesrat*; since that generally takes two to three years, it was decided to finance the first effort under Section 2.[19]

By mid-1987, a funding procedure had been worked out. Money from existing national research programs (on micro- and optoelectronics and on materials for future use) would be partly re-allocated to fund HTS research, and project applications could also be made according to routine procedures. A kind of incentive program, SUPRA2, was also drawn up to allocate a total of some $1.3 million to give HTS a smooth start. Clearly, this sum would not cover the costs of an entire research program; it was assumed that much of the necessary infrastructure was already in place and that ongoing projects could be partly re-oriented. It was expected that funding would go to innovative, suitably interdisciplinary proposals. At an ad hoc information meeting with no less than 150 participants, the funding agency announced these funding opportunities and invited all Swiss researchers to submit applications by November 1987 so that funds could be allocated by April 1988.

In contrast to Germany, the Paul Scherrer Institute in Zurich was the only non-university research institute involved in the national superconductivity effort. This research lab, the largest of its kind in Switzerland, had been created in January 1987 through the merger of the Swiss Institute for Nuclear Research (SIN) and the federal Institute for Reactor Research (EIR). Conventional superconductivity research was conducted at the Institute for Technical Physics, where some 20 researchers were involved in developing superconduc-

---

[18] See the "Protokoll der 368.-FR-Sitzung", June 2–3, 1987, Abt. II.
[19] At a very early stage an evaluation of the scientific and technological potential of HTS was carried out (Seeber, 1988).

tors for various technical purposes and, in the framework of a Euratom project, superconducting magnets for nuclear fusion. The group was experienced, large enough, and well enough equipped to shift to study the new ceramic materials when the HTS news arrived. Although the institute had good infrastructure, participation in the national program brought new specialized equipment and additional staff for the creation of an efficient and effective research group.

As with the conventional superconductor research program, applications remained the focus of HTS research. As the team leader explained, their research aims were not analysis and general investigation, but production with applications-specific properties. This meant, for example, the study of the specific orientation of crystals to increase their current-carrying capacity, or trying different methods of solidification from melts. Existing contacts with industry – including Asea Brown Boveri (ABB) – in the field of superconductivity were strengthened. To increase efficiency, a number of new, informal collaborations were set up with the Institute for Crystallography and the RCA labs in Zurich to characterize samples using X-ray and electron microscopy.

Swiss industry did not take the lead, but it did participate from the outset in national-level negotiations. But in contrast to Germany, little attention focused on establishing formal industry–university arrangements. Our Swiss interview partners said that the industry–university cooperation network already functions well and that industry was already deeply involved in decisions about the directions basic research should take in the universities and in other SNF-funded research labs Industrial presence not only proved fruitful for industry, it also provided impressive arguments for funding on the university level. It made the demand for funding seem credible and justifiable to a broader public.

ABB shows how different paths were taken in the case of HTS. In early 1987, when the company's scientists heard of the discovery, they set up a task force of 15 of the best researchers to collect information as a basis to decide the direction the company's research should take.

After a year of preliminary investigations, ABB decided that the potential of HTS warranted a research effort. A detailed program was formulated and a steering committee set up to ensure a balance between basic and applied research. The first phase of the project would last three years. This was an important commitment; it was ABB's response to the feeling that "in the past, new discoveries had always come from outside, from the universities", as we were told by our interview partner. The company had decided to reinforce its position.

For strategic reasons, ABB set up research projects at its labs in both

Germany and Switzerland. Government funding for industry is a normal procedure in Germany, but not in Switzerland. But here, as in many other respects, HTS became an exception. The expected amount of investment required to develop superconductivity led the Swiss to fear that only industrial giants could long remain in the race. So the SNF provided ABB money to carry out its detailed preliminary research program on HTS.

Spectrospin is an example of the smaller Swiss firms that joined the superconductivity *Arbeitskreis*. In the mid-1970s, Spectrospin began the production of superconducting magnet systems for nuclear magnetic resonance devices and later the construction of high-resolution mass spectrometers (90% of these products are sold to universities and the world chemical industry). Spectrospin's research concentrates on products; the superconducting wire components are imported from Germany or Japan, so the new HTS materials would be of concern only in the distant future.

Spectrospin was interested in the new field, but not actively engaged in research. It exerted a certain influence on the national level, acting as a consultant in formulating specifications for materials for use in devices for university research. Spectrospin also cooperates with the University of Geneva within the Eureka project on Chevrel phases, so it also had good access to information on developments in HTS. Finally, to compensate for its lack of actual research, Spectrospin had a network of consultants to provide it with up-to-date information.

Good industry–university relations played a central role in setting up Switzerland's HTS program so quickly, but several other reasons should also be mentioned. First is the good infrastructure in the universities. Many research groups could immediately use existing equipment and personnel to begin work on HTS, and could thus devote additional funds to improve rather than establish conditions. Second, a strong LTS community was optimally prepared to negotiate with ministries and funding agencies. In Switzerland, a number of researchers highly reputed in the field of conventional superconductivity started lobbying for national programs early. Swiss researchers could count on Alex Müller, who was drawn into the decision-making process as an expert. Third, policy-makers and scientists maintain very close relations in Switzerland. Swiss scientists had always been active participants in making decisions; keen on this research direction, they also functioned as expert consultants. Finally, Swiss funding bodies decided to interfere as little as possible in the organization of university research.

Ten project groups took part in the framework of the SUPRA2 program, and Section 2 allocated another $2.1 million to six other proposals approved through normal procedures. The national research programs noted above also re-allocated more than $700 thousand to HTS. Thus the total financial frame for HTS research was about $4.1 million, sufficient to finance 19 projects for up to three years.[20]

In sharp contrast to Austria, the Swiss decision to invest in HTS research was not only discussed in scientific circles but also received broad media coverage. An October 7 press release concerning HTS led to a widespread positive response in the national press. About 15 Swiss newspapers reported the decision on the very next day. That such an unbureaucratic way of funding research had been devised was seen as a "first" in Swiss science policy, and was considered important since Switzerland had "high research potential in low-temperature physics and superconductor technology."[21]

The Swiss demonstrated their commitment to HTS again in late 1988 when discussion started on the continuation of SUPRA2. Any follow-up program would be clearly structured, also aiming at the development of technological applications and commercially viable products in the mid-term – a commitment warmly welcomed by Swiss industry. The process of formulating a five-year national program was still regarded as far too slow and cumbersome. Instead, rapid unbureaucratic and non-routine decisions were made: HTS would continue to receive funds under Section 2, but now on the basis of a new five-year plan named SUPRA2, with a funding level of approximately $6.7 million. An ad hoc group including three advisors from Section 2, two external scientific advisors, and industry experts was set up to monitor the new program's progress. To promote long-term success, annual workshops would gather program participants and evaluate their scientific progress, so that the direction of research could be adjusted if and when necessary.

The basic research component of Switzerland's HTS program comprised three major themes: materials research (crystal chemistry, sample characterization); processing techniques (thin films, ribbon or multifilament superconductors, optimization of micro-structures); and the physics of the new superconductors (theoretical studies, basic properties, etc.). The deadline for applications to the program was mid-1989, and projects would be funded for eighteen months beginning in January 1990. About a quarter of a million dollars was allocated in the first round. Funding extensions would be granted following satisfactory evaluations.

In June 1990, only half a year after the start of SUPRA2+, the future of Swiss HTS research was already under discussion again. This time the Federal Council (the *Bundesrat*) had asked the funding agency to formulate a detailed

[20] The data are taken from "Förderung der Supraleitung durch den Nationalfonds", Nationalfonds Abteilung 1, April 1988.
[21] "Der Schweizerische Nationalfonds startet ein Spezialprogramm zur Erforschung neuer Supraleiter", press release, October 7, 1987.

plan for a national program that could take full advantage of the results and momentum gained with SUPRA2 and SUPRA2+. In this program (NFP30), "the acquired knowledge and competence have to be conserved and developed further, so that the necessary preconditions for utilization of these new superconductors by industry are fulfilled." In the future, priority would go to applied research; industry researchers, research labs, and especially small innovative firms were invited to join university scientists in participation. Research topics would include the development of new materials, production and characterization, improving sample preparation techniques, superconductor physics, and finally work toward applications and prototypes. The research projects began by mid-1992, with funding exceeding $8 million for a three-years period.

Austria

In the 1980s, basic research on conventional superconductivity in Austria was restricted to one well-established group at the *Atominstitut der Österreichischen Universitäten* and several smaller, more theoretically oriented teams. Two Austrian firms, Elin and Plansee, collaborated with a group at the Technical University of Vienna to produce wires and finished products from well-studied intermetallic superconducting materials. The Superconductivity Working Group (*Arbeitskreis Supraleitung*) coordinated activities on the national level, meeting regularly to discuss problems and exchange information. As noted earlier, collaboration with Switzerland had been established in conventional superconductivity in the framework of COST (European Cooperation in the Field of Science and Technology Research), which lasted for six years. Over a period of ten years, about $7 million from various sources was spent.

The complexity of the negotiations for Austria's HTS program is a result of the profound crisis Austrian universities find themselves in. Inadequate funding and staffing levels, structural problems in management, unsatisfactory career structures for researchers, and deteriorating infrastructure are the main problems. Awareness was growing that a university reform is needed to allow the Austrian research system to become competitive at the international level.[22] Cost effectiveness, quality assurance and control, and a stronger orientation toward the needs of society dominated the rhetoric and represented the guiding concepts for the future. But the situation will not change much, since the research budget allocated by the government is too small to cover

---

[22] A new law regulating the organization of the universities went into effect in Autumn 1994. Among its central ideas are the autonomy of universities, better management structures, and closer links to the market.

basic costs. Even if the universities find other funding sources, the administrative burden of preparing funding proposals and of managing the funds is increasing considerably. Other funding also means most of the money will be bound to specific projects, leaving little scope for research not oriented toward a mission. Infrastructure, office space, and laboratory facilities are often inadequate and appropriate research equipment is often lacking. Larger student bodies have not been accompanied by the necessary structural changes, nor by adequate increases in teaching staff. Senior staff turnover and mobility is low because virtually all positions become tenured after a few years; after that scientists are offered few other incentives to exert themselves in research.

This is the background before which the news of Müller's and Bednorz's discovery of high-temperature superconductivity reached Austrian researchers early in 1987. While some immediately started preparing their first samples to test the reported results, others discussed possible avenues of research and looked for allies. Only one researcher took immediate action. As early as May 1987, the leader of the superconductivity group at the *Atominstitut* submitted to the Ministry for Science and Research the extremely ambitious proposal to create a new Institute for Superconductivity Research. The proposal covered new buildings, equipment, and staff. He also recommended providing funds almost immediately to launch a research project to bridge the time until such an institute could be established.

The Ministry rejected the proposal, understandably enough considering the nature of Austria's science policy and the scope of national research. The creation of a new institute would have violated existing Austrian policy guidelines, which aimed to counteract the research landscape's increasing fragmentation into small units, a problem especially virulent in Austria. Additionally, superconductivity was regarded as too narrow a field to justify the continued commitment of money, manpower, and infrastructure such an institute represents. It wasn't certain that the new findings would prove sufficient to keep the new field scientifically and technologically interesting to ensure its long-term survival.

While the informal debate among scientists continued, the Vienna Science Fair in spring 1987 provided an ideal occasion to expand the discussion to include representatives of the Austrian Science Foundation (*Fonds zur Förderung der Wissenschaftlichen Forschung*, FWF), oriented toward basic research, and the Austrian Industrial Research Promotion Fund (*Forschungsförderungsfonds für die gewerbliche Wirtschaft*, FFF), which is applications-oriented. Scientists knew that, to keep pace with rapid international developments, they must quickly gain support from one of the research funds or directly from the Ministry and launch first research projects. Austrian industry was curious but unwilling to invest in HTS, whose applications were still remote. It soon emerged that the most likely source of funding for HTS research would be within the framework of the FWF.

At the time, the FWF had two main funding procedures. The first was the *Normalverfahren* (normal procedure), in which a research group submits a project proposal with a duration of up to two years. The second funding method was in the form of a *Forschungsschwerpunkt* (priority research program), which supported especially important research topics for a period of five years. This was intended to increase cooperation between research groups in Austria, to support interdisciplinarity, to avoid duplication and, to a degree, to channel competition. Ideally, such a program coordinates complementary individual projects on the scientific as well as on the administrative level. The only drawback of this funding procedure is that conception of a priority program requires applicants to invest considerable preparatory work in drawing up a stringent, long-term research agenda that finds the consensus of all participating groups.

When HTS was discovered, the funding agency for basic research was conducting an internal debate on how to strengthen the modest state of basic research in Austria. One outcome of the discussion was to create a new hybrid between the two-year normal project and the five-year priority program, which would be called a *Stimulationsprogramm*. It would have the same organizational structure as the priority programs, but funds would be allocated for only two years. If evaluation showed that the research was important enough, the program would then be transformed into a regular five-year *Schwerpunkt*.

Independently of the general discussion in the scientific community, four research groups in Graz quickly drew up a joint funding application to be submitted to the FWF in the form of a local priority program. They could have continued their solo run, but they were aware that the situation had meanwhile changed and that other groups in Vienna had also expressed their desire to begin HTS research. It was unlikely that the FWF would fund two parallel, large-scale undertakings, one in Graz and one in Vienna, so the interested groups tried to reach a consensus and formulate a nationwide proposal. This was intended to exert pressure on the funding agency and increase the chance of acceptance as a priority program with the five years of funding deemed essential.

The scientists' rhetoric to persuade the funding agency remained very conventional, however. They noted Austria's opportunity to take part in the international HTS effort and the fact that no single group had the manpower and infrastructure to compete internationally in such a fast-moving field; they suggested merging a number of activities in one national program to increase efficiency and foster interdisciplinarity; they called for a broad program to minimize the risk of choosing the wrong direction; they said if Austria made high-quality scientific contributions, it would gain access to international expertise and projects; and they noted that Austrian industry would later benefit from the experience and know-how gained in basic research. Virtually all the scientists involved were newcomers to superconductivity, and thus in no position to claim expertise.

While the researchers realized they had to collaborate to convince the funding body that a long-term program was essential, this first phase was characterized by a high degree of uncertainty, and formulating a common proposal turned out to be difficult. The researchers considered HTS so important that even a rough outline of general ideas of what they wanted ought to satisfy the funding agencies. The FWF rejected the first hastily compiled version on the grounds that the research objectives were too vague and that the programs showed no real cohesion. It made it clear that a collection of projects was not a sufficient basis for setting up a priority program and making a financial commitment for five years. But to demonstrate interest, by mid-1987 the FWF allocated $20 thousand for a preparatory conference and some travel funds. After this first disappointing confrontation with the reality of research funding, the Austrian researchers decided to make a second attempt and met again to negotiate a new research agenda in September 1987.

Meanwhile, one of the industrial enterprises (Elin) had taken the initiative and used its good contacts to organize a meeting at the Ministry for Science and Research. The meeting assembled the main actors – the researchers involved, representatives of the two funding agencies, the Council of Rectors (*Rektoren-konferenz*), industry, and the chamber of commerce (*Kammer der gewerblichen Wirtschaft*) and was intended as a forum to exchange information, declare areas of interest, and negotiate the possibilities of future HTS research on the national level. Although neither the government representatives nor industry made any clear commitments, they expressed the wish that any sustained research effort be coordinated in a form similar to the former *Arbeitskreis* for conventional superconductivity research.

Only two industrial enterprises wanted to engage in the HTS effort, and these were only slightly interested and only in the very early phase. The first was Elin Union AG, a state enterprise and part of Austrian Industries, which entered the field of applied superconductivity in the late 1970s. Its major effort was toward the production of conventional copper high-field magnets. In the 1980s, the focus shifted when a cooperation agreement was signed with CERN to investigate prototype Nb_3Sn magnets for a new generation of particle accelerators (LHC = large hadron collider), to be constructed in the 1990s.[23]

When Elin heard the news of superconducting materials with far higher critical temperatures in early 1987, the company initially regarded this as of sufficient scientific and technological interest to warrant a collective effort like that undertaken in conventional superconductivity. But although the company was active from the beginning, its own researchers were skeptical about the company's early participation. What complicated the situation was that Elin was just about to undergo a major restructuring, and the kind of basic research needed to develop HTS materials did not fit Elin's existing research agenda. The company preferred to concentrate on the development of marketable products such as high-field magnets without first investigating the basic materials. The time horizon for the development of HTS applications was at least ten years. Additionally, superconductivity research and magnet production were not a central effort at Elin. Large-scale investment in new equipment would have to be made to enable research meeting international standards. An Elin spokesman said this "would not have suited the structure of the firm." Moreover, Elin's superconductivity group was small, below the critical mass needed to do good research in more than one direction at once.

Thus although Elin triggered the early phase of discussion in the Austrian scientific community, it soon withdrew from the scene, restricting its research to aspects of LTS.

The other industry to get involved was Metallwerke Plansee AG. In contrast to Elin, Plansee is a small, dynamic enterprise with cooperative ties reaching as far as Japan. It entered the field of superconductivity in the late 1970s from the materials side, specializing in Nb_3Sn superconductors, a material ideal for high fields and high currents. Its superconductivity group worked on producing Nb_3Sn wires for use in magnets like Elin's and also on lead–molybdenum superconductors (Chevrel phases) within a Eureka project. Judging the company's position, the leader of Plansee's superconductivity group said it could "make wire, the wire is made into cables, the cable is wound into coils, and the coil will be put into some device." What kind of device and what market might exist for the device were unclear, and thus involved a high degree of risk. Small and ancillary industries like Plansee are highly dependent on the companies that use their products as components in larger devices.

With the discovery of HTS, the superconductivity group briefly changed its research agenda, slowing down its R&D on conventional superconductors.

---

[23] This illustrates the role other research fields played for the advancement of conventional superconductivity. In Germany (accelerators in Hamburg) and in the US (Superconducting Super Collider (SSC) and other accelerators), other research facilities are one of the most important markets for technological developments.

The group began with lanthanum oxides and then shifted to utilize its experience with ceramics by investigating the production of YBaCuO wires. This material's very low current-carrying capacity renders its utility questionable, even if wires can be formed. Also, due to their brittleness, YBaCuO wires easily lose their superconducting ability. Around the world, many research groups have tried to overcome these problems, for example by covering the superconducting powder with silver, but so far with limited success. Current-carrying capacities have been improved, but are still remote from the values needed for technological applications.

Given these problems, and knowing that German superconductivity specialists had set a time horizon of at least ten years for HTS applications, it was clear that, on its own, Plansee's two active researchers would not be able to keep up with international developments in the field. Plansee would need research staff and a relatively reliable return on any financial investment. Another possible way to take part in the new field would be to find an industrial partner with whom to collaborate. But this was difficult; most larger firms had already formed alliances. Our interviewee put it bluntly: it was as if within the first six months the "big electrical firms had already divided up Europe among themselves." With the growing recognition that thin films would be the first HTS utilization in microelectronics, Plansee, which had neither equipment nor experience to contribute, decided to leave the field again.

Like industry, non-university research laboratories played little role in shaping the national HTS program. The largest such lab, Seibersdorf, south of Vienna, had never made superconductivity a major research interest. Although materials science was one of its research areas, the lab decided not to make any in-house financial commitment to HTS research. The lab was also in the middle of a structural and financial crisis and had entered a phase of complete restructuring. It was clear that outside funding would be needed if research was to take this direction. Indeed, the initiative was taken on an individual level by one surface physics specialist and one technician from the Institute for Materials Research. They hoped to adapt an existing atomic beam spectrometer to produce thin layers of superconducting materials, and to investigate their surface properties. Management decided that the number of researchers and investments needed posed too great a risk in a field subject to such a high level of international competition as thin-film technology. Instead of rejecting the demand outright, however, a symbolic sum was allocated. It is no surprise that this group left HTS research after the first two-year program.

Summing up, the outcome of these prolonged negotiations within the Austrian scientific community and with the funding agency was a new proposal for a priority program on HTS, with a total financial frame of about $5.5

million for a period of five years. As we saw, few Austrian groups had worked on conventional superconductivity before the discovery by Müller and Bednorz. Some of the newcomers were specialists in measuring methods that could be applied to superconductors (magnetic measurements, spectroscopic analysis, etc.). Others were materials scientists familiar with oxide ceramics, making thin ceramic layers with chemical vapor deposition, or preparing samples at very high temperatures. All were convinced that their skills could contribute to an understanding of the new phenomenon. But in addition to scientific fascination for the new topic, opportunistic reasons surely played a role in wanting to join the research effort. Most Austrian researchers saw HTS as a welcome occasion to adjust the direction of their research and to improve their difficult financial position.

In spring 1988, a year after the first contacts, the funding agency decided to approve the project in principle but, to the dismay of the researchers, only in the form of a *Stimulationsprogramm*. Coordination of the various projects was assured, but with a reduced total budget of $2 million for a period of two years. The national funding agency felt that a planning horizon of five years was too remote in a field still moving so rapidly. Experienced groups would receive support to start work; other forms of funding might be forthcoming when research directions crystallized. The FWF argued that the Swiss National Science Foundation – often taken as a model for the Austrian science system – had also chosen to start with a program spanning only two years.

The FWF's main concern was still coordination on the national level. In its eyes, Austrian scientists preferred to cooperate with colleagues abroad than with other Austrians. It hoped the *Stimulationsprogramm* would counter this tendency and also create a critical intellectual mass, which the organization of Austrian universities did not otherwise foster. The two-year program would allow researchers to take part in the early phase, while more specific research directions could be negotiated at a more mature state. The field was changing so rapidly that plans were soon outdated and in need of revision anyway. A team of international referees who had been asked to evaluate the proposals supported the FWF's decision. Researchers had failed to persuade funders of the necessity of a long-term program.

While on paper the solution seemed satisfactory and comparable to other national programs, transforming the plan into research reality proved difficult. There were two levels of asynchronicity. First, though the overall decision had been made, some sub-projects within the *Stimulationsprogramm* could start immediately, while others still waited for formal approval. The second asynchronicity lay in coordinating staff and equipment. Groups that needed to purchase special equipment faced delays of several months. Thus they could engage staff but had no equipment in the first phase, thereby risking having no staff when the equipment arrived at the end of the project. This tends to be a problem for any short-term project, but it is exacerbated when infrastructure is poor.

In spring 1990, again following the Swiss model, the FWF evaluated the results of the *Stimulationsprogramm* and three related projects. Researchers were invited to present their work to three international referees, who would then give a report based on the publications and presentations. The results were disappointing as a whole. The work-sharing model, in which separate groups produced and tested superconducting materials, had worked only partially; using relatively big, expensive equipment on a time-sharing basis had not been a success; moreover too many avenues had been followed, producing solid expertise in none.

Despite the limited success of this first attempt, researchers expressed the wish to continue by submitting a request for another three years. All the tensions and difficulties encountered in the first phase hadn't discouraged them from preparing another joint proposal. The FWF eventually decided it would not fund a joint HTS program, but that individual projects should be submitted for evaluation. Some were rejected completely, others reduced in time or money. On the whole, the FWF felt that Austria's contribution to HTS in the first two years was not impressive enough to justify continued funding on a long-term basis. Instead of the $5 million requested in the second round, the FWF appropriated a total of $1.3 million to individual projects running up to two years.

4.3 Concluding observations: the role of policy

In retrospect, the complexities of the workings of the different national science policies and of the main actors defy any smooth summary. The ultimate outcome in each country depended very much on previously existing strengths and weaknesses. But systematic continuities were also interrupted by a marked reshuffling of scientific competence, interests, and alliances. Formal structures were important, but were overlapped and crossed by informal ones. Some of the initiatives preceding the national programs were more person-oriented than others, and key individuals were found in widely differing formal positions. Leadership and individuals played an important role everywhere. The ingredients of success in science policy-making seemed to be quick action to maintain momentum, the ability to build consensus by bringing in all relevant actors, helping them find a balance of interest and reasonable expectations for the future, and sound knowledge of the research groups' specific capabilities and potential.

Most scientists and research administrators agreed from the outset that HTS was not a simple or "successful" continuation of conventional LTS, but LTS researchers initially had a head start and a far better negotiating position than did newcomers. This was true for individual scientists asked by ministries, funding agencies, or other bodies to act as consultants as well as for those seizing the opportunity to push their own projects. It was also true for intact LTS networks as wholes. The overwhelming consensus was that experience in LTS was a definite asset, perhaps especially in view of the technological component.

On the other hand, Müller, the perovskite expert, had clearly shown that the new field would be open to many other disciplines as well. Especially in the early phase, tension between the old guard and the newcomers was noticeable at international conferences and within national research communities. Funding agencies in search of ostensibly expert advice were initially inclined to turn to scientists with LTS credentials. The latter naturally felt very excited and vindicated that "their" research field – even if no longer exclusively theirs – suddenly enjoyed such high prestige and interest. The relevant researchers who were not grounded in LTS were more dispersed, less visible, and less organized. In our sample of researchers, chemists and crystallographers were clearly less prepared than physicists to respond to the new opportunities.

Decision-makers had confidence in the LTS community, but the scientific community presented itself and its funding wishes to research administrators and funders differently in different countries. We have seen that almost anyone interested was invited to the first meetings, often held at the initiative of an individual in a research agency or of a member in a government ministry, in countries interested in industry participation. These initiatives were also a response to researchers' demands to "do something".

In Switzerland, the scientific community was represented by highly reputed scientists, and all decisions could be made on the basis of the country's considerable experience and long tradition in LTS. Moreover, Swiss researchers and decision-makers alike enjoyed a "Müller bonus", since the discoverer of the new oxides could act as an expert in guiding strategic decisions. Switzerland is small and very informal contacts prevail in its research system; it was soon clear who should be included in what capacity. The entire policy approach was informal and based on personal contact. No attempt was made to coordinate efforts, nor was it necessary to formalize industry links. Everyone seemed to agree that no special policy measures were required to ensure that cooperation would take place where necessary.

A larger country like Germany relied much more on formal decision-making procedures. From the beginning, the existing superconductivity community dominated, reinforced by advisors who had likewise cut their teeth on LTS. Over the years, researchers in universities, non-university institutes, and industrial labs had formed a stable constellation. Industry connections were reinforced through large programs and by including the large non-university research labs The hybrid community was longstanding and included civil servants and policy-makers in ministries and other funding bodies. For a time, a relatively closed circle had found new opportunities to enlarge its research activities – or so it seemed to the newcomers, mainly chemists and other material scientists who had never benefited from this stable funding situation.

Hoechst was the most spectacular industrial newcomer. Shrewd strategy allowed the company to enter the "old" network. Germany's new national research program in HTS closely resembled its *Verbundforschung* model, which brings industry and university research together, financed partly by industry, partly by government. The HTS program differed from others in that only universities could apply for funding in the first phase, while industry had to wait for a second round. The rationale for this exceptional procedure was to allow university groups a chance to catch up by better equipping their labs, so that they could be better partners for industry in the second phase.[24]

In Austria, the situation was much more confusing. Negotiations within the emerging hybrid community proved lengthy, conflict-ridden, and ultimately unsuccessful. A major problem was the ambiguous positions of industry and the funding agencies. The first large meeting at the national level was called by an industrial enterprise, but industry still showed no readiness to put its money where its mouth was. The funding agency for basic research acted similarly, showing initial interest but making no early commitments while it passively waited for the scientific community to act. The relationship between scientists and their funding agencies, including procedure-coordination on the national level, was very formal and rigid. Researchers felt the procedure was outdated and unsuited to local conditions. Much of the collaborative program worked out was doomed to fail from the start.

The way national programs are set up reflects the strengths of national

---

[24] After studying the German HTS research community, we do not agree with Dorothea Jansen's claim that the HTS "policy network" is simply "a transformation of the existing LTS policy network as a consequence of the intersection of actor strategies and new opportunity structures that are offered by scientific and technological change." The term "policy network" implies a much more regular and policy-oriented organizational and decision-making structure than was the case in setting up this national HTS research effort. In this context, policy simply means that additional funds had to be maintained and that allocative procedures had to be set up and implemented toward what were still vague research objectives. Germany's high degree of formal organization and the *Verbundforschung* model do suggest her interpretation. But our interview data shows that a new "hybrid community" – a configuration of actors from different backgrounds and operating at different levels – emerged in Germany as well (Jansen, 1991, 1994, 1995).

industries and their relationships to other actors in the research system. In the cases we assessed, the role of industry ranged from very active – and in the case of Hoechst, a substantially widened – involvement in the national research effort in Germany, to the Swiss case where collaboration with industry has a longstanding, good track record but functions quite efficiently on an informal basis. In Austria, industry never really got involved, a serious drawback in long-term research efforts, since it left university research groups with only their own intellectual resources and no chance of developing a rhetoric of economics to obtain funding. In the Netherlands, as well, smoothly functioning industry–university relations were judged as central to the setting up of the national program.

The consequences were a well-funded, long-term program in Germany, excellent and very flexible conditions in Switzerland, and real difficulties in setting up a clearly oriented program in Austria. Without discussing the US and Japanese examples in detail, these few European cases show the high degree to which industry continues to serve as a politically legitimizing factor for government funding of basic research, but perhaps even more importantly, how industry–university relations affect the more general directions national research programs take. This was evident in the guiding principles established for the second phase of these programs. Of course, the shadow industry casts on university basic research is much greater on a national than on the European level.

We expected HTS to serve as a focal point for tightening and redefining university–industry relations. Industry has often provided financial support to help establish centers within academia for basic research of potential commercial value, and governments support industry and universities that have these aims. There is a wide literature on the successes and failures of such efforts (Etzkowitz, 1990). Everyone agrees with such policies in principle, but difficulties are great in practice. Knowledge transfer often seems to work best when industry and university are in geographical proximity. Communication and mobility play a key role, as does the creation of a "critical mass" of related research programs, trained personnel, and adequate equipment. But planning a match of scientific minds and commercial aims is notoriously difficult when it comes to specific settings. Industries do not automatically utilize the fruits of research; nor are they always eager to produce knowledge themselves. Rather they want scientific knowledge "custom made" to fit into existing R&D plans and overall marketing strategies. Funding agencies are expected to create conditions allowing universities to produce such knowledge. Researchers wanting to demonstrate the seriousness of their claims to utility can hardly refuse efforts to strengthen university–industry relations. And industry that decides to enter a new research field should welcome the know-how offered by the universities.

In Switzerland, cooperation between industry and universities was smooth and informal. In addition, the same funding agency is responsible for basic and applied research, allowing flexibility and coordination. In Germany's initial phase, industry (except Hoechst) was more or less willing to let the universities have a first go at government funding. Of course, the HTS pie was four times as big as for LTS research. In the precompetitive stage practically everywhere in Europe, industry assumed that university researchers would provide it with knowledge rapidly and efficiently. Despite the ease with which basic research can be brought into industry, the loose nature of these contacts can also create weaknesses. European universities provide researchers with far less centralized aid in finding industrial partners than in the US, and researchers are not encouraged to take the initiative in making contacts. This is left completely to the individual research groups or departments. Governments, ministries, and some research councils play a more important role, but encounter obstacles of their own. European admiration for the Japanese model of organizing long-term, apparently stable, highly interactive industry–university relations reveals awareness that technological innovation is not so much a one-way "transfer" as an interplay reminiscent of a football team passing the ball back and forth among the players.

The national research programs set up in several countries did boost university–industry relations in the HTS field, but their effectiveness depended on a number of conditions. Where such links are traditionally strong, as in Germany and Switzerland, applications-oriented national research programs were set up with an ease bordering on routine. Perhaps the most interesting and novel phenomenon was that here, where improved university–industry cooperation is a goal in itself, the roles of funding agencies became interchangeable and the source of funding seemed to become irrelevant. Mixed funding was nothing new, but now ministries were prepared to finance basic research that traditionally fell under the competence of research councils and equivalent bodies. This is another sign of the increasingly blurred line between basic and applied research.

In the United States, on quite a different scale and under other institutional and economic conditions, the establishment of a government-sponsored university–industry consortium was hailed as the major outcome of the "wise men's" committee recommendations. It was seen as a test case for a new organizational form improving university–industry links at the level of basic research. So far, research progress has been satisfactory. The crucial role of graduate students has become visible; they function as go-betweens, shuttling between labs and their respective cultures. The consortium also revealed the need to include a small firm with sufficient courage to try to market the first HTS devices.

The sudden urgency to tap into knowledge not otherwise available is not the only reason why industry is becoming more dependent on university research. The many options for possible applications opened by basic science drive up the costs of basic research in industry, as many more alleys have to be followed in parallel. A much broader range of advances and knowledge produced in various disciplines must be monitored; this requires that research teams shift their configurations of interdisciplinary skills and expertise. The increased interdependence of technological fields such as telecommunications, computer technology, materials science, genetic engineering, and biotechnology amounts to a "technological convergence" (Mowery & Rosenberg, 1991: 213–14) greatly increasing the attraction and necessity of multidisciplinary research. Many recent experiments with new forms of R&D organization stem from industry's desire to contain the soaring costs involved. Pressure is mounting to narrow the focus of industry's in-house basic research. "Strategic" considerations in investment are increasing in importance. And consulting firms are eager to advise the managers responsible for research investment. Policy discussions have barely begun to address these developments' consequences for basic research and for the division of funding between private and public labora-tories.

National programs also reflect existing structures of research policy decision-making and the relative strength of organized research interests *vis-à-vis* the political establishment. HTS tested these structures for their adaptability to a new and unplanned situation, and for flexibility in permitting non-routine procedures for obtaining the desired results. Bureaucratic and other formal hurdles had to be overcome everywhere, even in Switzerland. Participants varied in clout; industry enjoyed a high degree of political legitimacy everywhere, while university researchers or research councils had to argue cleverly to convince funders. In all the countries we studied, existing science budgets were either insufficient or resisted shifts to cope with the new funding needs. Aside from the question of funding, HTS also showed how difficult it is to apply existing rules and procedures to a field changing so rapidly that any long-term proposals ran the risk of being obsolete before the funds arrived, and where no specific outcome of research could be reasonably predicted. Thus, funding agencies had to risk investing in directions whose promise might remain unfulfilled. Their consolation was that other funding bodies were in the same boat, and their precaution was to make funding short-term, usually for a two-year period. Results could then be evaluated, allowing a more selective procedure to be worked out. This approach resembles decision-making in industry, where basic research is conducted within strict temporal limits and then rigorously assessed before being continued or discontinued.

In the case of HTS, we observed the ascendancy of the funding agencies – research councils, governmental, quasi-governmental or inter-governmental bodies – that decide on national research programs and individual research projects in the name of peer review panels. We have seen how the establishment of HTS programs involved an extraordinary exchange of functions between research councils and governments. Did the source of funds affect university–industry coordination? Who should fine-tune the national programs; research councils with their affinity to university scientists, or ministries, which are closer to industry? If governments can step in at any time to finance basic research through their ministries and related agencies, what role is left for research councils?

Germany's Ministry of Research and Technology pushed the program toward applications. Industry was considered a legitimate partner from the start. To assure optimal coordination between academic research and industry, the ministry relied, as always, on the services of an intermediate body. In the initial phase, a very dynamic person was in charge who informally scrutinized all groups applying for funding, preselecting and shaping the national program. The research council, the DFG, followed its usual bottom-up approach, funding two special research programs but concentrating on basic research.

Switzerland's main research council, the Swiss National Science Foundation, circumvented formal procedures to set up hybrid structures responsive to what seemed the real needs of a program still requiring much basic research, but whose ultimate goal was applications. In Austria, the FWF funded basic research with its normal budget, but ventured to set up a new model, a stimulation program designed to "jump-start" the nation's researchers and leaving the directional decision for later. This new model included several coordinating mechanisms intended to optimize the use of personnel and other resources dispersed among various Austrian institutes. However, two things were not taken into consideration: the general conditions at Austrian universities; and the fact that locality plays an important role in knowledge transfer. Finally, in the Netherlands, joint funding was decided, most of it coming from the Ministry of Economic Affairs and far less from the basic research fund.

How much and at what levels does science policy matter? We have seen that preparedness was the key to joining the effort. And it arises from the history of a country's research organization and the life of its research groups. The imitation of organizational models developed elsewhere – like Austria's use of the Swiss "stimulation" idea – can be risky if other conditions are unfavorable. HTS was a test case for national science systems – what were the outcomes?

The creation of a research center for superconductivity was proposed in Germany and Austria, but the idea was not further pursued. In the case of Germany, it was judged too strong a pooling of resources and too high a risk of betting everything on one card. Austria saw a single research center as too costly and also as out of line with the general policy of strengthening cooperative ties among research units and institutes considered too small on their own. The UK science policy-makers, by contrast, were convinced of the value of centers of excellence. However, the example of Cambridge nicely demonstrated the fragility of this concept in a research environment structurally weak and financially ailing. Germany decided that interconnected local university research clusters were more promising, since they would bring together researchers from different disciplines and enhance local connections with industry.

Policy is fragile, subject to influence from rhetoric and political decisions. We expected to see the accelerating technoscience drift manifested in an altered set of criteria for selecting research for priority and funding, in order to concentrate resources. This was indeed what happened, fueled by a sophisticated discourse system of policy arguments moving from actual scientific practice through various institutional levels to the abstract level of international economic competitiveness. The discourse system is held together by a set of shared beliefs; but at each level, behavior, expectations, and rhetoric are framed and phrased to anticipate behavior and expectations on the next level. Scientists wanting to pursue "their science" present research proposals in a way implicitly or explicitly fulfilling the funding criteria they anticipate at the level of the funding agency: not only scientific excellence, but also potential technological benefits. Funding agencies apply the latter criterion in anticipation of the interests of the next higher level – ministries or other national political and administrative bodies. At the national policy level, criteria are widened again to include the strengthening of the country's position in international competition and the returns for national industry on world markets. Finally, at the international level, cooperative agreements emerge, couched in policy terms and other negotiations, tying science even more to other domains.

Configurations of actors in the policy space continuously form and transform themselves by cutting across the various levels. The configurations, which we term hybrid communities, are usually temporary, fluid, and overlapping. They are opportunistic in the sense of forming strategic alliances, since any action unfolding at any level stands in the context of legitimation at another level. Yet policy itself is based on scientific traditions and practice, industry's position in the world market, and national research budgets. Often goals, means, and legitimatory arguments do not match. Since policy is oriented to anticipate argument at the next-higher level to legitimate what happens at the lower level, it contains idealized descriptions and images of reality. Cooperation, the search for excellence, interdisciplinarity, or mobility always sound good in a policy paper, but are difficult to implement from above where their preconditions are not met in the everyday operations of the research system.

HTS displayed the limitations of and the opportunism involved in policy; these are inherent in the disparity between abstract ideals and a heterogeneous and concrete reality. Policy is usually formulated for the short- or medium-term, while building national research strength is a long-term effort. In the short period we observed, neither the size of the research effort nor its specific form significantly altered the overall balance of a country's research capacity. But at each level, almost all arguments invoked definite advantages for both the arguer and those to whom the argument was addressed.

A country's ability to keep up with worldwide research developments has become a precondition for exploiting new technological opportunities. This preparedness depends on university–industry relations, the capacity of funding agencies to respond quickly and flexibly, and the alertness of researchers. In HTS and other fields of basic science, success will ultimately be measured by the technological maturity achieved. Deeper scientific understanding of the phenomena is a goal found mostly on the lowest level of policy, that of project research.

At the middle policy level (that of national research policy), the relationships between university researchers, government labs and their equivalents, and industry assumes great importance. In our study, the strength of a country's industrial context definitely influenced the content and directions pursued by university researchers. The organizational structures, incentives, outlooks, and distribution of qualifications and skills still differ between the two realms; those countries with both a clear division of labor and strong cooperative ties have a great advantage. University researchers can clearly change their direction of research much more flexibly, while industry – once a decision has been made – is target-oriented, with high professional standards and clear deadlines. Inside the universities, research is more precarious; graduate students constitute an important but insecure investment of human capital and skills. On the national level, and supranationally in EU programs, efforts focus on promoting cooperation and on achieving a critical mass in research areas of strategic importance for industrial competitiveness.

At the highest level of policy discourse, the international situation is an incentive to intensify national policy efforts. The rapidly established national HTS research programs were widely regarded as policy test cases. They legitimated and reinforced efforts to keep up with world developments, increase organizational flexibility, and make it easier to shift budgets and people less bureaucratically. The internationalization of policy considerations and the inclusion of this level in the shared belief system thus led to a loosening of national control while paradoxically at the same time forcing national decision-makers to set up new and widened policy criteria and make them work.

Finally, let us turn to the symbolic or discursive side of science policy. We saw that all national HTS research programs were premised on claims about future achievements. Why did policy-makers agree to believe the promises offered them? The practice of science policy includes the beliefs, utterances, rhetoric, negotiations, and alliance-forming that necessarily accompany the establishment of national research programs. There is growing recognition of the importance of the "rhetoric" of science and technology and the policies implementing them. As public policy, the regulation of the research system is subject to symbolic politics – public exhortations and pronouncements of goals, visions, and expected results that mobilize the necessary human, financial, and political resources and create trust among those who carry out the policies, whether as decision-makers or as researchers.

In other words, science policy and research are dependent on belief. There can be no guarantee or hard evidence that a science policy responding to an unexpected discovery will identify the right research goals, let alone the sequence in which applications will ultimately diffuse through the various sectors of industry. But no policy is possible without this widely-shared belief in future returns for present investment and resource-mobilization. Empirical studies show that the key categories of science policy discourse, while socially constructed, tend to assume an air of objectivity. Reference and repetition in official policy documents and by persons of political or scientific authority lend an appearance of naturalness; the terms used become vehicles of imputed cause and likely effect. If they gain credibility, the social construct is reified and its contingent and socially negotiated nature forgotten (Cambrosio *et al.*, 1990; Wynne, 1992). The case of HTS illustrates that the importance of this symbolic aspect is increasing in the self-management of the research system.

In contrast to other empirically investigated cases, social negotiations in the field of HTS proceeded at a low level of conflict, and few alternatives were put forward to challenge prevailing policy assumptions. Everyone concerned wanted to establish HTS as a research field, and policy discourse was shaped accordingly. As Wynne suggested in another context (Wynne, 1992), interests and their symbolic representations mutually constituted each other. Interests converged to seize what was seen as a unique opportunity to obtain more research funds, while widely-shared beliefs upheld and enabled the necessary policy constructions. Any disputes or disagreements focused on which path to take to the shared goal.

Does that make HTS, with its modest achievements so far, primarily a case of successful collusion in science policy construction? Technological applications are still distant and there has been no major advance in the scientific understanding of the remaining and new puzzles of superconductivity. Was it all a waste of money, time, and energy – an instance of the media duping science policy-makers? Obviously, we don't think so. HTS provides a fascinating and illuminating glimpse into a research system undergoing changes on many levels and in interconnected, self-organizing processes. Science policy and its discourse can no longer be separated from the modes and content of research. The overall framework for science policy, while still provided by national research programs, is gaining in strength and becoming subject to new tensions through the growing climate of economic competitiveness in world markets. We observe a shrinking distance between basic research and expectations of uncertain future technological applications.

5

Science and the media: newspapers and their "HTS story"

One of the most striking features of the establishment of HTS as a research field is that – at least in the crucial first phase – negotiations expanded beyond the narrow realm of policy expertise into the public arena. The exceptional break-through was greeted by the media with great enthusiasm and was judged as newsworthy for a considerable time. Though not actively participating, the public came to play an important role as enthusiastic supporters or critical ob-servers of the field's evolution and as allies in developing an extensive rhetoric about the significance of the field's potential technological applications.

A trend toward increasing the degree of public staging of science and tech-nology issues has become more and more visible in the course of the second half of the 20th century. Dorothy Nelkin argues that this is closely linked to the fact that the societies we live in are increasingly shaped by science and technology. Members of these societies are thus:

continually confronted with choices that require some understanding of scientific evi-dence: whether to allow the construction of a nuclear power plant, or a toxic waste disposal dump .... The press should provide the information and the understanding that is necessary if people are to think critically about decisions affecting their lives
(Nelkin, 1987: 2).

Indeed, press reports described the intense excitement gripping scientists and science policy-makers alike in these first months following the breakthrough in HTS, and they elaborated spectacular scenarios of future applications. The me-dia eagerly seized on the notion of competition for technological and commer-cial advantage, playing up the potentially crucial role of the new materials in the development of an as yet unimaginable technology. Reporters soon drama-tized the competition among scientists, all of whom were eager to report they had found new superconducting materials with ever higher critical tempera-tures. In particular, the US media set the tone for what would be considered news, what content was worth reporting, and what attitude would be taken toward it. European science journalists, who generally take the US media as a primary source, followed, though with less excitement (Fayard, 1993).

Our look at the role played by the media, and through them by a wider public, in setting up the HTS research field proceeds in four steps.[1] First, we introduce our basic concepts on the communication between science and the broader public, present our sample of newspapers, and follow the chronology of the story as it evolved in the papers. Second, we discuss the factors that made it possible to construct an HTS story with broad appeal. Third, we show how science's entry into the public realm affected the behavior of researchers within the scientific community. Finally, we look at the different national contexts.

5.1 Science in the public space

In the following, we analyze, not a story of controversy or risk or fraud in science, but the media story of a fundamental breakthrough in basic science and of its attendant technological dreams – an example of the genre we call "scientific success stories". It introduced the lay public to a wealth of images of the frontiers of science. Readers were introduced to scientists given enough individual freedom and endowed with enough creativity and pioneering spirit to follow intuitions that contradicted conventional wisdom. At the same time, newspapers portrayed a new kind of researcher, able to adapt well to the changed conditions in the research environment. It offered a glimpse of a research area directly on the border between basic and applied research. And it offered a glimpse of science in the making, the negotiation of values, and other facets of the interaction between science, industry, politics, and a wider public.

The success story and the prevailing excitement were accompanied by some sceptical voices, in particular regarding the role the media came to play. Rustum Roy, Professor of Solid State and Geochemistry at Penn State University in Philadelphia, was among those who remained sober in the face of the orgy of enthusiasm. He contrasted the discovery of HTS with earlier discoveries in materials science, notably those of ferroelectricity in 1945 and of ferromagnetics in 1955. Roy noted that while the latter opened up viable programs of crystal chemical substitution, no superconductivity technology had so far taken off. He argued that, even if HTS reached the performance levels of competing materials, no "drop-in replacement" market existed, and that, even if HTS applications could move from liquid helium to the higher-

---

[1] Parts of this chapter are based on Felt (1992, 1993).

temperature range of liquid nitrogen, the savings would still be too small to fuel the growth of new industries.

At the time, Roy's scepticism was a relatively isolated position, at least publicly. He speaks of misjudgment by US science policy-makers, and blames it on "distortion by media hype". In his view, its irresponsible outcome was the result of "a joint venture between some (mainly university) scientists and breakthrough-hungry journalists". He judges the policies that put HTS on the funding map to be a clear aberration: "the pattern of HTS research activity is highly anomalous, while the pattern of advances does not deviate in any substantial way from other materials discoveries since World War II" (Roy, 1988: 28). But such harsh criticism begs the question: How can the media wield such a disproportionate influence, and what tempted university scientists to collude so eagerly with "breakthrough-hungry" journalists?

Clearly, the issues thus raised are not exclusively moral, nor can they be settled by appealing to the norms of correct behavior for scientists. Perhaps the seemingly "anomalous" pattern of research and scientists' behavior immediately after the discovery of HTS is becoming the "normal" pattern, in the sense that similar episodes can be expected to occur again in the future. Either conclusion underlines the obvious point that scientists do not work encapsulated in the scientific community and its value system. The products of their intellectual labor keenly interest not only their colleagues and a few industries, but also politicians, journalists, and the general public. Especially in the United States, researchers are increasingly aware that they work in a global context, and they act accordingly. They also know their work is not primarily valued for its stimulation of or satisfaction of the public's intellectual curiosity – though this remains an important aspect of what Dorothy Nelkin called "selling science to the public" – but rather for the technological and economic benefits it seems to promise.

Public controversies about technological risks have increased awareness that scientific and technological developments can lead to harmful as well as beneficial effects on society. A kind of hybrid public space has been created, where the media have come to play a new, and some would claim a disproportionate role. In the interactive game of public persuasion to obtain funds for research, to justify their expenditure, and to answer the increased demand for public accountability in science and technology, all participants have resorted to the media on one occasion or another. This new era of publicly announced and negotiated research preceded the advent of HTS. It began in 1980 at the latest, when the newly founded genetic engineering research firm, Genentech, put its stock on the market (Geiger, 1991). The media had become indispensable information providers for all concerned. Research news, of interest to potential investors and the business community at large, became a regular feature of the press.

Of course, informing was not the media's only interest. The media occupy a space which is:

> constantly being contested, which is subject to organizational and technological restructuring, to economic, cultural and political constraints and to changing professional practices. The changing contours of this space can lead to different patterns of domination and agenda setting ...
>
> *(Eldridge, 1993: 20).*

Especially when relevant to personal health or well-being or to the nation's economic prospects, dramatic announcements of the new and the unexpected draw readers, viewers, and listeners.

Indeed, the HTS story and its staging did not remain an isolated case, but was soon followed by the "cold fusion story", which developed along the same lines. While the latter was soon filed away as a scientific error unrelinquished by its authors, similarities remain in the ease with which the media picked up the story, aided and abetted by the scientists involved. "Science by press conference", though criticized and derided by those not directly involved, has its causes in the current structure of the science system (Lewenstein, 1995a).

Scientists' altered behavior toward the media, as the latter expanded their influence into the hybrid space of public interest and the acute demands for accountability, is only one of many signs of change in the relationship between science and society (Felt & Nowotny, 1993). HTS is one of many cases in which both researchers and funding agencies readily internalized and acted upon broad and diffuse societal expectations. The discovery of HTS was immediately defined as a spectacular breakthrough and, with rare exceptions, was regarded as a unique opportunity for research that would eventually lead to significant technological innovation. But why did the anticipation of largely unknown future technological benefits find expression in an unprecedented wave of public rhetoric and privately shared beliefs? Why were such beliefs sufficient to fuel the drive to set up research programs whose technological promises were flimsily based on what were still only research proposals in basic science? Another feature of the HTS story is thus the shared belief that the possible future technological benefits were sufficient reason to set up national research programs. The rhetoric developed to this end was extremely similar and simple all over the world; it nevertheless served to shape the cognitive orientation of the basic research programs. Even industrialists and the research directors of large corporations were ultimately persuaded to enter the new field and do their part in turning a technological dream into a distant but economically sound technological reality. How did the collusion of interests emerge and how was it enacted?

For the general public, popular print media like newspapers and magazines have become an essential source of information on issues involving science and technology. In particular, print media targeting the business community, investors, and political decision-makers regularly report science and technology news. But the print media's treatment of the HTS story took a new form. At key moments in the rapidly developing field, newspapers reported discoveries even before scientific publication; the story they constructed was sustained for a very long time; the press, and with it a broad public, became an arena of competition and were pushed into the role of spectators; and the way the story was told played a significant part in the mounting of the national and international research effort.

The media stories of HTS and, two years later, of cold fusion (Lewenstein, 1995a) show that the wider contexts surrounding science and technology have changed fundamentally and irreversibly. Science meets the public under radically altered conditions: trust and authority, support and cultural meaning no longer seem what they used to be. If the traditional "contract" binding society and science is no longer valid in its accustomed form, what has replaced it? Who negotiates the new forms of coexistence or cooperation at a time when public legitimation has gained importance? For the general public, the media have not only become an important actor in the negotiations between science and society, they are also the most visible arena of negotiation.

Communication between science and the public has long been naively understood in a two-stage linear model of scientists generating genuine knowledge and popularizers – journalists and scientists – writing a kind of simplified "translation" for the lay public, conceived as a passive recipient. This model was based on strong hierarchies. Scientific knowledge was thought to be clearly separable from and superior to popular knowledge, just as scientists held expert status and were clearly distinguishable from lay people. Information flowed in only one direction, from the science system to the lay public. Communication was thus mainly reduced to a translation process, and much effort was concentrated on discussing aspects of language and the possibilities linked to it. At the same time, these models contained a strong image of enlightenment, the idea that more scientific information would automatically lead to a more positive attitude towards everything having to do with science and technology. Since the 1970s, however, social studies of science as well as some efforts in communication sciences have moved away from this dominant image and started to deconstruct the process of dissemination of scientific information.

We now understand that when the mass media provides news about science, it is fabricated by the very process of reporting. The media never straightforwardly provide a correct, if simplified account of reality, but try to stage a drama calculated to appeal to a broad audience. Stories have come to be seen as "the product of cultural resources and active negotiations" (Tuchman, 1978: 5). The media choose between different kinds of information, highlighting some features rather than others, and also choose a particular language. Metaphors are invented and pictures added as the narrative unfolds. But above all, these stories actively define norms and deviations. Thus we believe that the discourse between science and the public, which takes place in an often very indirect and not easily discernible way, should be understood as a negotiation of the meaning and relevance of scientific information, rather than that it can be located somewhere between distortion and adequately accurate simplification.[2]

5.2. Analyzing our newspaper sample

Ours is a qualitative analysis, supported by quantitative information, of coverage of HTS in newspapers in three German-speaking countries (Austria, Germany, and Switzerland) and the United States. We chose nationally distributed periodicals that regularly report on science and technology. We did not systematically investigate magazines specialized in popularizing science, but we did have about 50 articles from *Bild der Wissenschaft*, *Science News*, *Discover*, and other popular science magazines, which served as background to the study of the articles in newspapers. Since our interest was not quantitative, we restricted ourselves to a modest sample (see Table 1). Our 316 articles included 151 from the US, 83 from the pre-unification Federal Republic of Germany, 56 from Switzerland, and 26 from Austria. We analyzed the general presentation, the key words in the headlines, the placing of the article in the periodical, the key issues, the metaphors used, the main actors, the sources of information attributed in the story, and many other qualitative features.

Frequency analysis of the articles

A cumulative frequency distribution of articles published from January 1987 to December 1989 (see Figure 1) sheds light on the press's construction of the HTS story in different national contexts and on the media's perception of what constituted major events. Correlating the frequency curve with actual scientific,

[2] For a detailed discussion of the relation between science and the public see also Hilgartner (1990); Shinn & Whitley (1985); Cloître & Shinn (1986).

Table 1. *Number of articles published in the four countries from December 1986 to December 1989*

Country	Newspaper	Abb.	Total number of articles analyzed
United States	New York Times (d)[a]	NYT	
	Wall Street Journal (d)	WJ	
	Newsweek (w)	NW	
	Time Magazine (w)	TM	
			151
Federal Republic of Germany	Frankfurter Allgemeine Zeitung (d)	FAZ	
	Süddeutsche Zeitung (d)	SDZ	
	Frankfurter Rundschau (d)	FR	
	Die Zeit (w)	DZ	
	Der Spiegel (w)	SP	
			83
Switzerland	Basler Zeitung (d)	BZ	
	Neue Zürcher Zeitung (d)	NZZ	
	Tagesanzeiger (d)	TA	
			56
Austria	Die Presse (d)	DP	
	Der Standard (d)	DS	
	Salzburger Nachrichten (d)	SN	
	Trend (m)	TR	
			26

[a]d = daily, w = weekly, m = monthly (d)

technological, and political events in the field shows that "news" from the frontiers of research depends not only on scientific developments, but also on political declarations, economic expectations, the awarding of prizes, and other events. This nicely demonstrates the importance of being able to link scientific events to other important parts of social life to make them newsworthy. Indeed, this clearly supports the arguments put forward by Cloître and Shinn.

In the case of popularization of science the language, the reasoning, and the images do not manage to elucidate the phenomenon, but quite to the contrary there is a tendency to create a conceptual incomprehension. [...] Popularization constitutes thus not an efficient instrument for the transmission of a better knowledge about the physical world. Its force and its pertinence lie in the links it establishes between a scientific subject and the social sphere

(Cloître & Shinn, 1986).

Media attention in all four countries broke down into six key events or periods (Table 2).

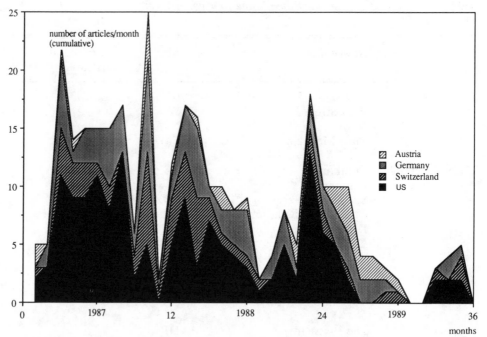

Figure 1. Articles dealing with HTS in the newspapers mentioned in Table 1.

The number of articles published in each country varied widely, as did their distribution over time in the three European countries (see Figure 2a–d). In the topics chosen, we found notable differences among the three European countries, as well as between them as a group and the United States. Only in October 1987 was there a synchronous European peak in reporting on the Nobel Prize award to Müller and Bednorz. This was clearly a "European event" reported less profusely in the US. In Austria, the daily press does not accord scientific issues a high profile and science journalism is hardly an established profession; this may explain the slow and stilted response to HTS, especially compared to the United States or Germany. Researchers in the field wrote all three of the long, rather detailed articles printed in one paper. They clearly bear the mark of natural scientists and are less journalistic in style. Press reports were much more frequent in Switzerland than in Austria, at least during peak periods, which largely coincided with the peaks in the US media. Later the number of reports tapered off, with no clear pattern emerging. Germany's press reporting differed markedly from that of Austria or Switzerland, showing regular oscillations and not tapering off until Spring 1989, thus following developments for a longer period. German reports concentrated on issues related to HTS technology and applications, probably reflecting the interests of German industry.

Table 2. *The six key events or periods*

March 1987	Meeting of the American Physical Society; about 3000 researches meet to discuss HTS; the press labels it "The Woodstock of Physics"; this event draws worldwide attention. (One year has passed since the discovery by Müller and Bednorz). US labs find new materials with critical temperatures even higher the Zurich oxides
April–mid-July 1987	Construction of the scenario of competition between the US and Japan for the market for high-tech. products
Late July-Aug. 1987	HTS becomes a political issue: President Reagan calls for a conference to discuss potential applications and excludes foreign researchers; 11-point superconductivity initiative to support HTS research
October 1987	Müller and Bednorz receive the Nobel Prize
After Jan. 1988	New HTS materials (Bi- and Tl-compounds) are discovered with a T_C as high as 125 K
December 1988	Discussion of patents and applications and, especially in the US, of the construction and financing of the superconducting supercollider

5.3 The public history of HTS

The discovery of HTS was a European event. The initial discoverers were Swiss and German, and the laboratory was situated in Rüschlikon, near Zurich. But the first press reports on the event appeared in the United States; this greatly influenced the reconstruction of events and set the tone for later perception. Reporting on a press conference held by Paul Chu of the University of Houston as well as on information received from AT&T Bell Labs on New Year's Eve, 1986, the front page of the *New York Times* trumpeted "2 Groups Report A Breakthrough in Field of Electrical Conductivity". The article restricted itself to generalities, providing few scientific details but hailing the success of US research and immediately drawing an enthusiastic picture of the discovery's possible implications – especially technological implications. Paul Chu is quoted as saying that if critical temperature could be raised to 77 K, then "superconducting technology would no longer be restricted to some so-called high-technology exotic, expensive technology.... We can also use it for large-scale applications" (*NYT*, December 31, 1986). In contrast to a later tendency, the central actors described in this article were institutions – the

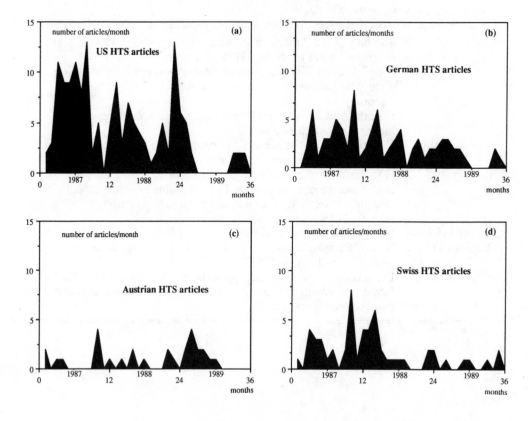

Figure 2. The number of articles published per month in each of four countries.

University of Houston and AT&T Bell Labs. – with the researchers taking secondary roles. The fundamental work of Müller and Bednorz was not mentioned until well into the second part of the article, on page 13, and the two researchers were cast in the role of a trigger, rather than as intellectual leaders in the new field.

Public interest in HTS was still considered low in early 1987, and no follow-up reports were printed in the media we investigated. This first *New York Times* article took a special position as the primary source for later reports in the European media which, in this early phase, did not question its strong bias toward successful US laboratories. Even the Swiss press gave credence to this version, not investigating further or constructing their own story about what had been done at the IBM lab in Rüschlikon. At this stage, few European reporters noted the central role of Müller and Bednorz, who would receive the Nobel Prize only nine months later. In this phase, European reports generally did not spotlight individual researchers, except in one article that mentioned Chu. Nor did they dramatize the technological promise of the discovery.

A second major article on HTS appeared again on the front page of the *New York Times* in mid-February 1987, once more in response to a press conference called by Paul Chu. While scientists, especially at US universities, often announce their findings in the general press, Chu's conference was unusual because he called it before publishing his results in a recognized journal. This act already mentioned had several purposes, but above all it served to claim priority in discovering a new superconducting material before others could do so.

Meanwhile, thousands of researchers around the world had entered the field, creating a climate of intense competition – a fact eagerly taken up by the press. Local newspapers like the *Houston Chronicle* published reports under sensational headlines, such as "Discovery may earn billions and Nobel for UH [the University of Houston]". The national orientation of the story increased, with Paul Chu as the hero. All the major elements needed to construct an appealing media story were already present in a less elaborated form: visions of future applications; lower costs due to cheaper coolants; the staging of the beginning worldwide "race with few parallels in the history of applied physics" (*NYT*, February 16, 1987). The HTS discoveries that now followed each other in rapid succession provided a "fairytale come true", with glory heaped on the "scientific world" and the promise of technological wonders and economic benefit for all.

March 1987

Reports on HTS took a new turn with the American Physical Society meeting in March 1987, which was dubbed the "Woodstock of Physics". Colorful descriptions of this meeting filled the US press: "The doors opened early Wednesday evening to a roar, a blur of color and a stampeding abandonment of professorial dignity. Within three minutes, the crowd had filled all 1200 seats, and nearly 1000 more physicists jammed the aisles and pressed against the walls. Outside, hundreds more strained to get in" (*NYT*, March 20, 1987). "I think our lives have changed," the much-quoted phrase of a Bell Labs physicist who attended, aptly expresses the prevailing mood. The US press published a strongly personality-dominated series of articles on HTS that presented all the exciting facets of the achievement.

From the outset, reports on HTS were characterized by a naive enthusiasm for the technological promise of the new materials. Over the three-year period we observed, five areas of application were popular from the beginning: transportation systems using levitated trains; faster and smaller computers using superconducting components; energy systems utilizing loss-free storage and faster transmission systems; medical diagnostics equipment using MRI (magnetic resonance imaging); and "big science" facilities such as particle accelerators or fusion experiments incorporating superconducting magnets. Reports also speculated on possible military uses of HTS, including reconnaissance, radar, and other applications yet to be developed in the context of the Strategic Defense Initiative (SDI) program. Media in the United States mirrored the enthusiasm and optimistic assessments common within the scientific community in this early period. Potential benefits to the general population were highlighted to legitimate the allocation of public funds for research.

The articles employed a rhetoric reminiscent of earlier periods of technological optimism to conjure up a vision of technological utopia. Indeed, the HTS story confirms the arguments of Dorothy Nelkin: "In the 1980s the technological enthusiasm of the 1960s has been born again though somewhat tempered by the continued fear of risk. The idea of progress has been resurrected as innovation; the celebration has re-emerged as a high technology promotion. The old cliché of breakthrough has reappeared" (Nelkin, 1987: 10). In her investigations of press coverage of science and technology in the United States, particularly in regard to the changing perception of science as reflected in newspaper articles, Nelkin uncovered what she terms "cyclic trends". While the 1960s saw a period of scientific and technological "breakthroughs" and "revolutions", in the late 1960s and 1970s, wonder at the marvels of science and technology gave way to concern about environmental and social risks. Journalists turned from conquests to consequences, from the celebration of progress to critical reflection about the problems arising from technological change. The HTS story seems to be an indication of a return to a more optimistic phase of the cycle.

The language of the press articles emphasized novelty, the adjective "new" appearing in almost all headlines. Other eye-catchers included "gains", "major advances", or underscored saved money ("without energy loss", "cheap superconductors"). Some headlines focused on record critical temperatures, vying with each other in the use of superlatives ("higher", "newest", "fastest", "the most important breakthrough since the transistor", etc.). No serious obstacles to any of the applications were discerned in this public space.

In the United States, interest in the story was heightened by concern about competition with Japan for high-tech markets. The story evolved in an oscillating tension between pride in American research and pessimism about American ability to turn research results into viable technologies. A typical article lamented that HTS would "once again show the competitive drive and speed with which Japan can seize Western science," only to end by quoting a Japanese scientist regretting that his country was only "following in the footsteps of the United States. Here again the originality is coming from the West. We have a measure of sadness about this" (*WJ*, March 20, 1987).

April–mid-July 1987

After March 1987, US reports on HTS retained their strong national orientation, focusing on Chu and other individual researchers – an approach they would continue throughout the time of our study. The German and Swiss press began to construct their own story, while Austria's media did not yet respond at all. The content and several key features still resembled US reporting, but Müller and Bednorz and the IBM lab at Rüschlikon were cast in the leading roles. The BaLaCuO superconductors they had discovered were labeled the "Zurich oxides", indicating that the mainstream of the story was being reorganized and that the European aspect of the story was increasingly present. Müller and Bednorz were given clear credit for having "broken a psychological barrier" and having made the whole development possible.

The US press continued to cover HTS, but widened the range of topics discussed. The articles printed in April were still extremely optimistic, ranging from utopian speculations to interest in details of developments in applications. Articles focused on the progress achieved since March, including advances in the production of thin films, single crystals, and a number of plasma-spray technologies. The construction of the first small test devices elicited great interest, and the possibility that the Nobel Prize would go to a leading HTS scientist was discussed several times.

The waning of the euphoria allowed first doubts to be voiced in headlines including expressions like "daunting problems" or "a series of hurdles" (*WJ*, May 5, 1987). For the first time, the general media questioned whether critical temperature was a sufficient criterion for a "good" superconductor – the assumption that had been central to the staging of a competitive race. Sobered headlines started to alternate with continued sanguine reports such as "Superconductors cross big barrier", which mentioned a "rapid series of optimistic findings".

The story regained momentum with an increasing number of reports about new materials at ever higher critical temperatures. The *New York Times* headline on June 5, 1987 read that no less than "4 Labs Achieve New Gains on Conducting Electricity" with a subheading that "Resistance is lost at close to room temperature in their tests". The media picked up any news of this sort, even without learning or reporting the details. At the end of June, however, when confirmation of these announcements was still lacking, explicit scepticism set in. It was realized that "as reports of ambiguous findings multiply, just what counts as proof of a temperature breakthrough in superconductivity is becoming hazy" (*WJ*, July 1, 1987).

Several other issues also enter the stage. Discussion began on potential investors moving into the field and their chances of making profits. Awareness also grew that a theoretical explanation was still lacking, and would be necessary for both scientific and technological progress. Attention also turned toward the funding and future institutional status of HTS in the US, and also, for the first time, toward the way the science system functions. In particular, the new role of the media was underlined by the issue of "science by press conference".

The rhetoric of the German articles was far less sophisticated than that of the US press. It concentrated on scientific or technical details and often provided a historical review of the discovery, but addressed fewer wider questions. Many articles underscored their own "scientific character" by quoting or reproducing original graphs from scientific journals like *Nature* or *New Scientist*. In the German press, the progress of the story was less closely linked to current reports on discoveries than in the US, often lagging a week or more behind events. Science policy issues were seldom subjected to public discussion and the general tone of reports remained optimistic for the time being. Typical headlines were "Electrical perpetual motion in the deep-freeze", "Loss-free current slowly approaching", or "Soon components made of high-temperature superconductors". Only a few articles discussed these dreams of superconductivity critically. The central, optimistic message was that researchers would turn their findings into viable technologies in "the not-too-distant future". When discussing the eventual awarding of the Nobel Prize, it was taken for granted that Müller and Bednorz would be the recipients.

Swiss press reports in this period kept as close as possible to the actual research developments, with as few allusions and speculations as possible. This attitude was also reflected in their choice of illustrations, such as graphs, sketches of crystal structures, and photographs of technical apparatus, rather than the mystical images of levitated magnets or the exciting high-speed trains that figured prominently in US reporting. The Swiss press also reported on progress in conventional superconductors, quite apart from the issue of HTS. It was assumed that time would pass before HTS would mature to the application stage.

The first Austrian report on HTS and the only one in this period gave a historical account that broke off in January 1987! All the crucial later discoveries went unmentioned (*SN*, April 4, 1987).

Late July–August 1987

In late July and early August, HTS entered the sphere of politics as the media reached the climax of its focus on economic competition. Newspapers reported on President Reagan's federal conference as "responding to the perception of a strong challenge from Japan". The conference brought together government labs, industry, and university researchers to discuss possible applications of HTS, but excluded all foreign scientists. President Reagan underlined the importance he attached to the field by announcing his 11-Point Superconductivity Initiative.

Austria did not report on these developments, but Swiss and German newspapers criticized this nationalistic gesture, seen as a dangerous precedent in limiting the free flow of information that would change the character and function of scientific work. Headlines ran "Worry About American Secretiveness: Superconductivity as First Serious Case of Encapsulation" and "Superconductor Conference in the US: Foreigners Excluded". Some reports expressed anger and disappointment about this unfriendly act, while others cast doubt on the realism of Reagan's technological visions (*NZZ*, July 31, 1987).

The absence of genuine scientific news other than regular reports on signs of superconductivity near room temperature was compensated, on the one hand, with more comprehensive review articles and, on the other hand, by discussing new features of the organization of research that had become apparent in HTS. An article titled "Scientific Saga", for example, raised the question of what amount of "crazy thinking and luck" was needed to advance in science. Fear was growing that the current organization of research, requiring concrete proposals for research planned well in advance, left little room for unconventional ideas. Under the subheading "Keeping secrets from bosses", the article further speculated that a research proposal for work like Müller's and Bednorz's would never have been approved (*WJ*, August 19, 1987).

Turning to the topic of applications and obstacles to their development, much space was devoted to presenting and discussing the different standpoints of industrial companies. The range of perceived potential applications remained unchanged, and rhetoric turned on "vast corporate profits", "multi-billion dollar industries", and "commercial competition". References to the market relevance of superconducting materials can be seen as calls for increased industrial investment. Analogies of the transistor and past success stories like the laser and the integrated circuit were often invoked. This technological/economic angle was much more common in the United States than in Europe, in line with the political importance President Reagan placed on improving the nation's competitive position. By referring explicitly to interested firms like Japan's Mitsubishi, Toshiba, Sumitimo Electric Industries Ltd., and Nippon Telegraph & Telephone Corporation, Germany's Siemens, Dornier, Interatom, and Vakuumschmelze, and US companies like Texas Instruments, this facet of the story gained a more realistic touch and gained credibility.

At the same time, headlines were also becoming more realistic; for example, the August 2, 1987 *New York Times* announced "A Sober Pause In Superconductor Race: Applicability Is Not Just Around The Corner". Thus the ambivalence toward projections of further scientific and technological developments had become an integral part of the staging of HTS.

An important focus of the articles was still on the rapidity of developments in the field, and metaphors of a race were omnipresent: "the challenge", "competition", "winning", and "supremacy". Terms like "the frontiers of science", "burst", "race", "rush", and "hunt" conveyed the impression that competition was still fierce. Indeed, international competition, in particular between the US and Japan, became the leading topic.

October 1987

After September 1987, the media reported fewer new results in superconductivity, but took up other themes, such as the frustrating attempts to produce thin films, Japanese patent applications, crystal growth, and computer technology. The lull ended suddenly, however, with the October announcement that the 1987 Nobel Prize for physics had been awarded to Müller and Bednorz. Naturally, this triggered a peak period for HTS reports in the European press. Swiss reports highlighted the special role of the IBM lab at Rüschlikon and some Swiss papers issued special supplements. Headlines such as "The IBM research laboratory at Rüschlikon again in the glamour of a Nobel Prize" addressed a question that was in the minds of both scientists and policy-makers: Why did this lab produce such outstanding scientific results and thus win the Nobel Prize in two consecutive years? What were the working conditions that led to this highly innovative work? Thus the personal careers of Müller and Bednorz were described at length, the long and often difficult path they had to take before making their discovery was traced, and in particular the concept of research freedom in the framework of IBM was brought to the fore.

The Austrian press awoke to the euphoria of the event, devoting four relatively short reports and a review article to it. While expressing admiration for the Swiss lab and the two outstanding scientists, the articles also grasped the opportunity to stress that, in the high-tech sector, Austria could not even compete with small countries like Switzerland.

The German media played up themes similar to the Swiss, and also praised the Nobel Committee's courage in awarding the Prize so soon after the discovery of HTS. Of course they also celebrated Bednorz as a national hero.

The US reported on the award in less detail, but the Laureates themselves were portrayed in a *New York Times* article. The scientific community in the United States was unhappy about the early announcement of the prize, believing the committee had thus excluded potentially "worthy recipients" who entered the field late. They particularly lamented that their national hero Paul Chu had been left out. The award provided the occasion to go into the subject's history once more.

The Nobel award strongly influenced HTS research and also caused a shift in the media's reporting. While the media still propagated images of speed and dynamism ("race", "running", "at high speeds", "a torrent of research", "avalanche", "frantic search", "blow after blow", "rapid succession") and still presented room-temperature superconductivity as the primary objective, the situation had definitely calmed. That the highest scientific honor had been granted for the discovery of HTS, however, served to stabilize the field and established a basic consensus on the importance of the findings of Müller and Bednorz.

After January 1988

A short period of scepticism, patience, and restraint then followed. On January 18, 1988, the *Wall Street Journal* published a very reserved account of developments in HTS under the headline "Despite Advances, Doubt Surrounds Race for Room Temperature SC". It began with the statement that "scientists have become increasingly sceptical about reports of superconductivity at high temperatures" and continued that, for almost a year, "the thermometer has been stuck at about 93 K". It suggested that progress depended on "another quantum leap with different materials".

At the end of January, the mood shifted again with the discovery of bismuth superconductors in a Japanese lab and then in US labs. Regional US newspapers reacted immediately; national ones were slower, indicating that interest in the field had been sated. The number of articles on HTS consequently increased, but they did not reach levels of euphoria comparable to those a year earlier. Press optimism focused on the cheapness and availability of the raw materials needed for the new Bi superconductors.

By now, the term "breakthrough" was associated with past achievements, though the discovery of thallium superconductors soon thereafter showed that the field was far from being closed. The general prospects for applications received less and less attention, presumably because they were reserved for the distant future and because journalists assumed readers were already largely familiar with this aspect of the subject. Concrete examples, like the construction of the first electromotor using HTS materials, commanded more interest. The motif of US–Japanese competition persisted and focused now on commercial applications and patent announcements, reflected in headlines like: "Commercializing Superconductivity: If we lose the race, blame industry" (*NYT*, March 27, 1988).

Priority disputes, now become a public topic, also enhanced HTS's newsworthiness in early 1988. Two cases are worthy of mention. Swiss papers reported a controversy over Tetsuya Ogushi of Kagoshima University, hailed by one Swiss researcher as the "first discoverer" of HTS (*NZZ*, February 3, 1988). An emotional exchange of opinions within the Swiss scientific community ended with the repudiation of the claim. Ogushi may have produced superconducting compounds before Müller and Bednorz, but he failed to recognize them as such.

Similarly, Chu and Wu were embroiled in a dispute over property rights. Some US journalists jumped at the chance to report on a personal disagreement. Chu had sold the rights to his "1–2–3 superconductor" to DuPont without giving any share of the potential profits to Wu, his collaborator and the co-author of the initial publication.

From April to December 1988, no new scientific findings were announced, and articles on HTS grew shorter, appeared less frequently and did not make it to the front pages anymore. Apart from some general themes, such as US–Japanese competition which was elaborated in considerable detail in December 1988, a great number of articles were devoted to the US decision to construct the superconducting supercollider (SSC), the world's largest and most powerful particle accelerator; it was speculated that it would incorporate magnets made of the new superconductors. (The SSC project has meanwhile been abandoned.)

The number and length of reports on superconductivity continued tapering off in 1989, indicating media saturation in the absence of new developments. The number of references to Chu, Müller, and Bednorz fell to an all-time low, and the now familiar list of potential applications was repeated. Only in the

United States did the dramatic appeal of competition with Japan, theoretical controversies, and science policy questions still draw readers.

5.4 The ingredients of a scientific success story and their national variations

Having described the publicly constructed history of HTS as a research field, we now draw the threads together. What were the main ingredients in constructing the scientific success story of high-temperature superconductivity in the media? How did they vary with the national context? Why didn't the three German-speaking countries develop the degree of euphoria observable in the United States? What journalistic traditions affect the news-worthiness of scientific information? How are scientific events linked with other conditions to achieve public visibility for purposes of policy negotiations?

In our sample of articles, we identified six elements that made the story worth continuing over a longer period of time. All six hold for the US media, thus explaining their sustained account; but in the German-speaking countries, some were missing or only mentioned in passing, so that the story developed in other ways.

Economically and politically relevant science

The first and clearest thread running through the whole HTS story in US newspapers was the interaction between science and the political and economic systems (Dickson, 1984). These three subsystems of society are always interdependent, but HTS made these links particularly visible. Changes in the way the news of HTS was constructed is a sign that some fundamental changes are taking place in the science system. As a researcher expressed it in one of the newspaper articles, "all of the [research] processes are being accelerated... We're considering questions related to technology at a much earlier time than I've ever heard of" (*N YT*, April 9, 1987). These observations fit well into the ongoing discussion of whether the science system has entered a transition phase (Rip, 1990; Ziman, 1990, 1994) in which basic research is more closely tied to applied technologies and is more oriented toward the market than before, with new forms of accountability – in short, of increasing influence from external criteria on decisions in the science system. If scientists are more aware of external criteria and rewards and put more effort into creating their research environment and "selling" technological promise, then the media and the general public have a more important role.

The motif of economic competition dominated the field from the beginning, especially as soon as it was realized that the initial technological optimism exceeded research realities and that commercial applications were not "just around the corner". But the HTS story had developed its own dynamics. It stood for American worry about Japanese commercial competition in the markets for high-tech products. As early as March 1987, the *Wall Street Journal* printed an article titled "Japan Is Racing To Commercialize New Superconductors". It and others depicted a keenly competitive Japanese research effort in the field. The line "when it comes time to make something out of it, the Japanese will have the upper hand" clearly shows the tone of such articles (*WJ*, March 20, 1987).

In the 1960s, there was a similar peak in the number of scientific reports on LTS. At that time, Japan was not considered a major competitor of the US; rather, readers were told that LTS "was one of the research fronts that received great attention in both the United States and the Soviet Union, both of whom were engaged in a race for supremacy in science and technology ...." (Derian, 1990). In the United States, the presence of an opponent/competitor, whether real or imaginary, seems an indispensable ingredient in arguing and winning the case for federal support of the national research system. Thus, although the technological race with the Soviet Union is over, "luckily" another competitor has been found.

This tendency climaxed in late July and early August 1987, when HTS moved onto the political stage. Newspapers reported extensively about President Reagan's "11-point Superconductivity Initiative, the points ranging from a new advisory panel of 'wise men' to propose legislation that would strengthen patent protection and relax antitrust laws," as "responding to the perception of a strong challenge from Japan...." The *New York Times* front page article headlined "Reagan, Citing Foreign Challenge, Outlines Superconductivity Plans" quoted the President as noting, "Science tells us that the breakthroughs in superconductivity bring us to the threshold of a new age." Superconductivity was no longer the affair of a small number of scientists, but had taken a central role in political and public discourse. The promises of HTS were said to be enormous, including "a reduced dependence on foreign oil, a cleaner environment, and a stronger national economy", all topics eagerly seized by the media because the public was easily sensitized to them. The situation was further dramatized by an appeal from the National Science Foundation to America's industry to begin thinking in time frames greater than two years or risk that "the others will reap the benefit of the insight that our laboratories have produced" (*NYT*, July 29, 1987).

The theme of competition seems to be the thread running through US newspapers throughout the whole period. From early 1988 on, reporting was continuous, even on the development of small technical devices, such as an electromotor using HTS materials. Competitive fears and accusations against domestic industry appeared regularly in forms like:

Japanese corporations have been pouring millions into research and engineering and are already boldly announcing new superconducting products. ... These are just a few of the initial commercial products announced ... and they are arriving much sooner than any American experts have predicted

(NYT, March 27, 1988).

Press reports in the German-speaking countries took a different tack. Enthusiastic about the scientific and technical progress reflected in headlines like "The most important discovery since the transistor", they still made no firm promises of applications in the near future. In the early phase, HTS was seen as "For the moment only in the lab." (*FAZ*, March 13, 1987). After Reagan's press conference in July 1987, the tenor of the German press shifted slightly, devoting some attention to economic competition, technological relevance, and first applications, even if they do not leave the laboratory stage. The Swiss press expressed fascination – "Superconductivity: With Zurich Oxides to the Current Wonderland?" – but clearly delineated the obstacles to commercial realization (*BZ*, March 18, 1987). The only early-period report in Austria did not allude to applications at all (*DP*, March 21–2, 1987).

"Good old little science"

In the initial phase, the HTS discovery story held particular appeal because the field was believed to require virtually no funding and few infrastructure investments. It was perceived as linked to the ingenuity of individual researchers. At a time when science has been growing ever more complex and demanding enormous investments of money, manpower, large and sophisticated equipment, schedules spanning decades, and the collaboration of several hundred scientists, the prospect of an inexpensive but potentially lucrative science was welcome. HTS arrived at a moment when US politicians, scientists, and citizens were discussing whether to invest billions of dollars in the construction of a superconducting supercollider – a machine that would allow the US to regain the lead in the field of elementary particle physics – and when many fields of science were suffering from short funds. HTS could be presented romantically, evoking the "golden age" of science. Individual scientists and their intellectual capacities, intuition, creativity, and perseverance were portrayed as the keys to scientific progress.

An example of this is the March 31, 1988 *Wall Street Journal* article headlined "His Was the Little Laboratory that Could", which told the story of Allen Hermann, the leader of the team that found the thallium superconductors.

> Mr. Hermann's team entered the superconductivity race late and had little money for laboratory equipment. It succeeded through pluck, luck, and sweat, illustrating a major reason why superconductor research draws so much excitement: Even small labs make major contributions
>
> *(WJ, March 31, 1988).*

Many of these very individualistic stories also paint a picture of the scientific endeavor very different from and in some ways contradicting the image of an expensive, extensive, and bureaucratic science system mentioned above. Planning, strategy, and conceptualization take a back seat to a vocabulary of chance, luck, alchemy. The new materials are even described as "fruit salads" accepting any ingredient (*WJ*, February 10, 1987).

Another aspect should be noted. Steven Shapin's article on the public understanding of science argues that modern society is distinguished by a radical disjunction between residences and work sites. The latter are specialized in space and in knowledge. This separation is also true for science, creating a "fundamental problem for the place of science in a society with democratic pretensions and aspirations" (Shapin 1990, 1992). Indeed, science is an occupation usually not very visible to the lay public. Contact with science is generally restricted to school, the mass media, and entertainment. Even here, because experiments have grown more complex, it has become much more difficult to demonstrate science. But the media eagerly noted that HTS was different. In January 1988, the *New York Times* ran an article "Superconductors move from the laboratory to the classroom" (*NYT*, January 12, 1988). Initial experiments were easily repeated "even in school classes" and at science fairs. The image of a superconducting disc floating over a magnet seen through a mist of evaporating liquid nitrogen had become a symbol for the mystical attraction of the new discovery. HTS had brought the general public back in direct contact with science again. As one article noted, "if anything is capable of rejuvenating student interest in science, it's superconductivity" (*NYT*, January 24, 1988).

Hero scientists

The third major ingredient in the story was the "hero scientist". The US media selected a number of researchers such as Paul Chu to accompany the public throughout the story. The dramatic appeal of the story was greatly enhanced by the tendency to personify research and equate the worker with the work. Many of the public's strongest beliefs about science derive from its image of scientists, "principally because those beliefs include assumptions about how the appearance, personality, and intellect of scientists relate to the importance and consequences of their work" (LaFollette, 1990: 66). In a detailed study of the public images of scientists as portrayed between 1910 and 1955 in US popular science magazines, Marcel LaFollette has shown that personalizing the description of research makes abstract ideas more comprehensible to the lay public. Many of the articles we studied underlined the special role of individual researchers, the human links between the scientific and the public worlds. Scientists are expected to act in ways beneficial to both realms; as normal citizens and at the same time experts, they are expected to share their knowledge with society.

But the "public performance" of scientists in US newspaper accounts of the HTS story differed considerably from that in the European countries. US media usually attributed scientific advances to particular scientists or teams, and individual researchers dominated the accounts. In European reports, with few exceptions, scientists – even "local heroes" like Müller and Bednorz – were cast in rather impersonal roles. They were termed discoverers and experts, but the social aspects of science and scientists' role as human actors was downplayed in the European HTS story. Reports in the German-speaking countries still preserve the traditional image of science as a mere knowledge-producing activity subject to intra-scientific criteria.

US reporters, by contrast, adopted a variety of fanciful images: "Science's men of La Mancha, dreaming the impossible dream"; scientists "plunging toward the impossible"; and "trailblazers" in the scientific jungle. Throughout the period we studied, scientists were cast in the role of heroes, combining optimism about a better future with insatiable curiosity, restlessness, and the drive and ability to explore new paths. Though concern was expressed about their publication behavior, the complaints were muted and avoided direct criticism of individuals.

Hard work was another of their heroic attributes in the US HTS story. We find "research teams working around the clock, seven days a week . . . .. Sleeping in shifts, they cooked their meals in a tiny kitchenette while their latest batch of experimental ceramic pellets baked in the lab's kiln." Readers were invited to "watch" Paul Chu:

in his tiny office at the University of Houston, wearing, as usual, a long white lab coat over his jeans, his head framed by a rough helmet of black hair. He is speaking excitedly into the telephone, using one of the increasingly common languages of US physics – scientific Chinese, every fourth word an English technical term. In the laboratory a few feet to the left, the red glow of a furnace is visible.

Elsewhere, we find Tanaka described as "a chain-smoking, coffee-gulping 36-year-old with a mid-five-figure salary" or Müller as a "brainy" and "gentlemanly Swiss physicist", while "Batlogg's life has been one part of adrenaline and two parts obsession". We learn that Hermann "often seems to boil over with excitement about his work", but also "that he plays the trombone well enough to have performed with Ella Fitzgerald …."

Along with descriptions of the scientists' working situations – and sometimes their private situations as well – we also find some more mystical attributes. Bernd Matthias, a leading figure in the superconductivity community, was said to possess a "magnetic influence" and to draw his "best ideas from dreams". Paul Chu was described as imbued with "deep experience" and as having an "intimate relationship to his materials". The same was maintained about Müller and his "intuitive feeling about oxides", while Bednorz was character-ized as "a whiz at the lab bench".

Despite the national differences in presenting science to the lay public, there has also been a global change. Scientists are no longer portrayed solely in terms of their intellectual excitement, the thrill of discovery, or the satisfaction of acquiring new knowledge. The heroes of the HTS story are not depicted as altruists. Journalists came to accept that the "noble" activities of researchers around the world had a real-world background of intense competitive pressure. The competition was not only for prestige, reputation, research funds, or other intra-scientific values; scientists were also under increasing pressure to prove that their avenue of research would contribute to their nation's ability to compete economically.

Chu's portrayal is a good example of the new image of scientists for the 1990s: the affirmation of competitiveness, full cognizance and willingness to enter a race whose stakes are no longer solely personal prestige and scientific honor, but also national commercial and technological prowess.

Staging a race

We have seen that the press portrayed the HTS story as a race. This staging was greatly eased by the fact that the complex scientific competition could be put in terms simple enough that spectators with little scientific background could still see who was ahead: the critical temperature became the central (and linear) parameter. The motto was: the higher the temperature, the better the scientists, the lab., and the nation hosting the lab. For scientists, on the other hand, it soon became clear that critical temperature was not the only criterion for high-quality superconductors, and the technoscientific soon took a different path than the media story. While the media concentrated on the easily discerned dimension of temperature and later also on the development of minor applications, the technical story followed material parameters like current-carrying capacity or reproducibility, useful for technological applications.

In a sense, the scientific race was over when Müller and Bednorz won the Nobel Prize in 1987, not even a year after the community started working in the field. Speed had increased not only in research, but also in the granting of recognition and rewards. With the ultimate goal already achieved, the race for recognition gave way to a commercial competition for patents fiercer than ever. This race continues, with the US and Japan the major contenders.

Science without risk

A feature distinguishing the HTS story from other discourse on science and technology is that it dispensed with warnings of potential hazards, technological catastrophes, large-scale environmental damage, and individual exposure to risk. HTS still appears free of the perceived public risk that has grown to be part of science reporting. HTS was thus suited to revive technological optimism. It was linked to innovative technologies that promise to alleviate problems in energy supply and transmission or to usher in a technological utopia. Media coverage thus recalled a time prior to the energy crisis and pessimistic scenarios of the "limits to growth".

Science vs politics

Bored by Iranamok? Had enough of 'insider-trading' investigations? Take heart. While politicians dance atop the heads of legal pins, science has just presented two of the most dazzling discoveries of our time – superconductivity and a grand and puzzling supernova. Researchers seeking superconductivity have been science's men of La Mancha dreaming an impossible dream

(WJ, March 30, 1987).

This was the start of an article published in the *Wall Street Journal* on March 30, 1987. Comments in German papers from the summer of the same year join it in exemplifying another facet of the HTS story: science as distracter from an unsatisfactory political situation; science as displayer of the marvels of nature;

science as an ideal basis for stable political ties; science as impartial mediator in the international arena.

A look at the press reports from the various countries reveals wide variations in the balance between these different key elements. Why was this the case? What effect did these emphases have on the duration and intensity of public interest?

Although the study of HTS involved researchers from many fields, including chemistry and crystallography, the majority belonged to the solid state physics community – an area that was under severe financial pressure. With the support of the media, scientists hoped to gain public visibility, to demonstrate the scientific and technological relevance of their research, and so to put pressure on policy-makers and funding agencies. The aim was not only to give a research field new impetus, but more importantly to secure long-term funding. But telling the HTS story fulfilled aims other than securing money for the field in general. It was also an ideal case to demonstrate the unpredictability of scientific results, and thus to plead against the increasing mission-orientation of research and for more leeway in choosing avenues of research, with less constant pressure for predictable results. And the newly gained visibility of individual researchers or research labs was a valuable bargaining chip in competing with others for resources on the national level.

For science policy-makers and politicians, staging this kind of story was also advantageous. Science had moved into the limelight with a positive image, creating a competitive climate that provided a greater range of policy options and a firmer basis for legitimizing funding decisions. Further, science – and in the case of HTS, "little science" – seemed to promise a strengthening of the national economic position. With such intense media coverage, policy-makers were under pressure to act, but they could also expect that their decision to fund research would be seen as justified.

Finally, the media were able to continue presenting an exciting and colorful picture of scientific research for quite some time. The story offered suspense, action, fascinating visions, and was not spoiled by images of risks.

What led to the clear national differences in the way the story was told? The US media story embraced all six facets outlined above, thus permitting a relatively high continuity in the telling. Beyond that, in the HTS field, science and technology had become inseparable, which was useful in mobilizing public interest. This confirms LaFollette's remarks that the American public tends to

see 'science' as instrumentally valuable, just one among many human activities that can improve the quality of life. . . . This practical view of science might well help to explain the conflation of science and technology so prominent in American popular culture: science appears as a more esoteric but nonetheless practical initiator of technology

(LaFollette, 1990: 176–7).

The three German-speaking countries varied among themselves, as well. Austria had no continuous HTS story at all. First, the country lacks a culture of science journalism, and second, there was no one interested in colluding to tell the story. While most US universities are aware that they need the support of the general public to get public money, in Austria science funding is not particularly subject to public accountability. The relationship between industry and academia is extremely weak, and Austria's industry developed little interest in cooperative research. So it is hardly surprising that discussion of HTS's applicability or economic and political relevance was meager. The press' perceived cultural duty was its main interest in reporting scientific developments to the lay public. So articles appeared occasionally, but no story or coherent message was constructed.

German reporting focused on economic considerations and science policy themes, thus somewhat resembling media attention in the US. With Siemens, Hoechst, and Daimler Benz eagerly involved in HTS research and the Federal Ministry for Science and Technology funding a large-scale national program, the media had enough material to fuel its activity. But while US media homed in on international competition (especially *vis-à-vis* Japan) and the fruitful links between a nation's science and economic systems, German articles stuck more to technical details and possible technological applications. In January 1988, three out of four articles dealt with a prototype electromotor, ignoring major scientific advances. As in the US, technological dreams and fiction were sold alongside more realistic appraisals of the situation. The American preoccupation with hero scientists had its counterpart in the German treatment of Bednorz as Nobel Prize winner.

Swiss newspapers responded very differently. Although the discovery was made in a lab near Zurich and Müller was Swiss, no reports were published in the Swiss press until 1987. The news had to come from the United States. Only later did Swiss science journalists begin giving the story their own twist, centered around local heroes. Regular science reporting is customary in Switzerland; and articles tend to use a relatively technical style, indulging in less futuristic speculation. Perhaps this is partly explained by the fact that the Swiss superconductivity community was already well established in the national context, and thus felt no need to mobilize the public for their cause.

The thread running through the entire HTS story is the link posited between basic science and economic competitiveness. This is especially interesting in light of the recently increasing questioning of the assumption that a good science base helps a nation succeed economically (Brooks, 1987). In a climate of fierce economic competition, stagnating budgets, and a public attuned to negative consequences of science and technology, selling science is no longer a luxury for the scientific community, but both an obligation and a strategic necessity.

5.5 Science in the marketplace: scientific competition and the media

Competition in science is nothing new. Whether on the individual, institutional, national, or international level, it has always been a major driving force for advance. Metaphors and analogies, as we have seen in an earlier chapter, have compared it to a sporting contest or have interpreted it in more economically oriented terms. But the forms of competition, the prizes scientists compete for, and the way this is made visible to a larger public prove to be strongly correlated with the social context and the way research is organized. In this sense, we argue that continually increasing public attention to science and technology and the semi-autonomous role of the media in reporting on science are a major intrusion upon the scientific realm.

Earlier parts of this chapter traced the construction of the HTS story over time and its appeal to the public; we now turn to the individual researchers. What drove scientists to move beyond the intra-scientific realm, its norms, criteria, and established system of reward and recognition, into the hybrid space we have described, where the public plays an important role and the media provide the interface and arena of negotiations? To what degree do the media and the public affect scientists' behavior in a highly competitive situation? How do scientists adapt their strategies, and what does that imply for scientific research on the global level?

Scientists who act as media sources can be rewarded for it in at least five ways: personal satisfaction, and recognition by the public, employers, politicians, and peers (Friedman *et al.*, 1986: 10). Our interview partners – who were all from European research labs – rarely address this subject, and then not explicitly; and most of them saw their role as a passive one, if any. Few thought about the necessity of and means of interacting with a larger public, and few actively interacted with the media. However, our press material and the public discussions reported in science magazines allow some conclusions.

Another remarkable feature of the establishment of HTS as a research field was what came to be known as "science by press conference". Scientists began announcing their results to the general press before or even instead of submitting them for refereed publication in professional scientific journals. Marcel LaFollette's study of science magazines shows that scientists have always used the mass media for their purposes:

Especially in the early part of the century, scientists from every discipline used popular magazines for a different type of reputation building, as public forums from which to answer critics of their work. Researchers in a new discipline or proponents of a new theory, discouraged by criticism or rejection by colleagues, sometimes felt compelled to circumvent traditional channels for scientific communication; when that occurred, they often used popular magazines as pulpits from which to evangelize the scientific community while simultaneously arguing for public acceptance (and possibly funding) of their work

(LaFollette, 1990: 53).

At first glance, the situation with HTS seems a little different; it was no longer mavericks, but well-established and even high-ranking mainstream scientists, whose work was not seriously threatened with rejection, who began using these extra-scientific channels of communication.

One reason for this appears to be that, in an exceptionally competitive situation, scientists felt trapped by their own rules and norms and by the existing structures of reward and credibility. The public arena seems to have given them both a feeling of greater freedom and the possibility of openly formulating original ideas opposed to mainstream scientific belief. They thus could trigger discussion otherwise not possible. Further, it seemed to open new avenues to building reputation. Jacobi and Schiele have brought this nicely to the point:

It is of secondary importance if the information processed [in the course of the popularization effort is wrong or right or up to what degree it is wrong or right, compared to the fact of the existence of this discourse and the interpretative framework this offers to the practitioners

(Jacobi & Schiele, 1988: 14).

Under normal circumstances, when a research field is stable and institutionalized, the lay press is assigned the task of "diffusion of results" and "popularization", but must be legitimated first within the scientific community. Only when unusual and unexpected events occur, as in the race to produce new HTS materials, does the clear boundary between communication within and beyond scientific circles blur (Shinn & Whitley, 1985).

Press reports on the HTS story served at least two functions. First, moving into the public space from the more rigidly codified framework of the scientific sphere allowed the development of a sophisticated and speculative rhetoric. The potential for technological applications could be emphasized beyond any realistic appraisal, elaborating on the importance of the breakthrough for industry, the national economy, and the whole of society. Second, the press intervened in an unprecedented manner in the "making of scientific facts" and in making or breaking reputations. The public not only witnessed advances, but was also called upon to referee priority disputes. The scientific community felt this effect strongly, since it violated firmly established and accepted norms of scientific publishing conduct. The normal procedure is to submit a paper to a journal, undergo peer review, and wait – usually several months – for publication. In the euphoric phase immediately after the discovery of HTS, it became customary for scientists to establish priority claims in the lay press before submitting papers to specialized journals. This practice could be observed again shortly after in the cold fusion story.

This shift toward more opportunistic behavior can be traced to a number of factors in no way unique to the discovery of HTS. First, normal peer review broke down under the avalanche of new findings. Not only was the procedure considered too slow to fairly document priority claims, fear also grew that reviewers would not maintain confidentiality. The potential commercial value of patents and intellectual property rights to scientific discoveries is tremendous, and a corresponding temptation. In the United States, universities have moved to secure control of the commercial exploitation of their research findings.

This phenomenon is well illustrated by Paul Chu, whose publication behavior was discussed in a long article in the New York *Time Magazine* headed "In the Trenches of Science". The title implied that the scientific community was "at war" and needed better protection for intellectual property. Chu's paper, submitted to *Physical Review Letters*, contained "typographical errors" that were corrected very shortly before publication; many observers think the errors were inserted deliberately, to protect his work. Perhaps Chu's mistrust was justified; if there were no leaks, it is difficult to explain why so many groups proceeded to use the material "erroneously" mentioned in the original paper. This affair triggered a wide debate about ensuring priority, especially in areas where prestigious awards such as the Nobel Prize and lucrative patents are at stake. But at the same time, the science system was seen as robust:

When these conflicts recede from memory, a story will remain of scientific discovery in its purest form. The heroes will be a few obsessive physicists driven to understand the strange, shimmering, electronic qualities of crystalline matter, and who took a path that their colleagues either scorned or overlooked

(*TM, August 16, 1987*).

Scientists have thus recognized that they can use the media to protect their own interests. This move results from increasing strains in their relations with traditional sources of funding and from new patterns of accountability for the social impacts of science and technology. The media's response to HTS reveals not only the changed relationship between science and industry, but also a move to open the core of the research system (where judgments about the significance and quality of research are reserved to peers) toward the periphery (where the media hold sway). The shift may be temporary, before a return to normality, but it does occur, and with the full collusion of the researchers.

A July 1, 1987 *Wall Street Journal* article discussed these changes at length:

... the recent rash of reports on high-temperature superconductivity annoys some researchers, who feel that publicity-conscious colleagues are breaking the rules by announcing their findings before the results are reviewed by technical journals. That enables a researcher to establish a claim of being the first to make a finding, yet without divulging details that would usually appear in a scientific journal and potentially aid competitors.

At the same time, the pace of superconductor discoveries in recent months has overwhelmed the journals, complicating efforts by scientists to publicize their findings and duplicate colleagues' results. Physics journals 'are about two months behind....'

The charge was picked up a few days later in *Newsweek* (July 6, 1987), which also discussed at length the problem of the field's advances outstripping the capacity of journals to publish them. News was breaking every week, but articles sometimes took a year or more to appear in print. This cast great doubt on the effectiveness of peer review in such an exceptional situation.

Further, instead of relying on normal publication channels, researchers were sending papers directly to each other – "science by fax". Müller said he had a foot-high stack of unread papers on his desk. The frenzy also led to charges of sloppiness. "In this field you don't retract anything," a scientist was quoted in *Newsweek* "you just stop claiming it." In the case of HTS, this phenomenon appears to have been largely limited to the United States; at least we find no trace of a similar discussion in other countries. This fascination of flash over substance seems to reflect "the problem with the way physics is done in America, where some people are declared to be stars and some subjects are declared to be hot by a very small group of people" (*NYT*, April 5, 1987).

The above must be seen in the context of a global inflation of the amount of information the public receives on political, social, economic, and increasingly on scientific and technological issues.

How does scientific competition change when the scientific lab extends to a wider space including the public, and what role do the media play in this? Various competitive strategies entered into complex interaction when the HTS research field was forming. Scientists compete individually for reputation, positions, money, prizes (including, in this case, the Nobel Prize), etc. But, as team members of universities or research labs., they also compete against other institutions for reputation, good students, and financing. And as specialists in a field, they vie with other research fields for prestige, perceived "usefulness" (in terms of applications), and thus funds. Finally, nations compete with each other. While these different levels of competition surely preceded the new, more active role of the media and the public, the terms of competition seem to have changed.

Press reports not only document priority claims and describe new results, they also provide scientists with three new negotiation assets: public authority, increased public interest in the discipline, and contribution to national goals.

Scientists use the media to gain public authority as a key person, an expert in the field in the eyes of the wider public. As long as a field restricts itself to communication and rewards within the scientific community, the rules for gaining recognition are clearly defined. Reputation usually grows slowly, over a period of several years and, once attained, enjoys a high degree of stability. When the media become an essential element in the environment of a research field, they have the power to attribute credit and make or destroy reputations. Public authority becomes almost as important as peer recognition, since ranking high in the public eye is capital when negotiating for public funds and, jobs. But this asset is unstable and potentially counterproductive.

The second value negotiated is the degree of public interest in the field of research. This depends on the benefits a wider public expects from it. From the beginning, media interest in HTS focused on potential applications. Indeed, more than 75% of all articles discussed applicability. The "public value" attributed to a branch of research is not so much the result of a realistic economic and technological appraisal of the situation as of the common strategic act of what Dorothy Nelkin (1987) has called "selling science".

Related to this is the third value, the possibility that research will contribute to national goals – which may include military security, medical advances, or economic gains. HTS research embraced all of these. It was thus an ideal case for bodies funding research. In the United States, federal funding for research is closely tied to perceived national goals; this gives more weight to the voice of the media than in other countries.

The media thus played an important role in structuring scientific competition in HTS. They created a new kind of hybrid space in which science is negotiated, and where the value of science is judged in accordance with prevailing societal visions rather than by intra-scientific criteria. To adapt to these changes in their environment, scientists are increasingly obliged to play this new game – at least in situations of high competitive pressure.

6

The innovation machinery of science:
the case of HTS

Many observers agree that science is currently passing through a period of dramatic transformation. At the end of his lucid analysis of science in a dynamic steady state, John Ziman concludes that there is no way back to the traditional habits of managing research, but there is also no obvious path forward to a cultural plateau of comparable stability.

The new structures that are emerging are not the products of a gentle process of evolution: they are being shaped very roughly by a dynamic balance between external forces exerted by society at large and internal pressures intrinsic to science itself

(Ziman, 1994: 250).

We believe that the emergence of HTS sheds light on what these forces are and how they interact. In the beginning of this book, we compared the effects of the discovery of HTS on the research system to a building tested by being subjected to a transient load which reveals otherwise hidden strengths and weaknesses. Indeed, HTS can be seen as a case that shows how complex and fluid the present situation has become. Researchers can no longer expect to find an environment hospitable to their work, but are compelled to create one. We have seen that it takes extraordinary effort, time, and energy to set up the conditions under which research programs can run for a predictable period. Such efforts are no longer external to, but have become an integral feature of scientists' work. Nor are they limited to the small research group institutionally at home at the university.

The situation is also extremely fluid on the level of policy-making. Nowhere is science policy firmly in place; rather, it is continuously shaped and reshaped under the impact of new and old requirements, such as priority setting, selectivity, strategic thinking in research, and the future exploitation of scientific knowledge produced today. There are two main reasons for these policy imperatives. One is that the global context of the production of scientific and technological knowledge has thoroughly changed. No longer does a small group of highly industrialized Western countries hold a virtual monopoly. Even basic science has been drawn into the orbit, by now seemingly inescapable, of enhancing international economic competitiveness. The second reason is the increasing expectation that basic research will prove useful. Scientific knowledge production itself has moved into different contexts of application, resulting in a widening of the selection criteria for funding. Scientific excellence is now one criteria among several, to be negotiated on different policy levels.

The case of HTS allows us to observe these different and changing features interacting in novel configurations and to analyze their effects. Taken by themselves, none of the contributing features is new or unusual. But taken together, they provide a fascinating glimpse into the present working of the science system as an innovation machinery. Our study has shown how its national variations function after an innovative discovery has been made. HTS tests how prepared science and research systems are to respond. The responses themselves depend on a number of interacting opportunities and constraints, which we have analyzed in some depth.

We examine the historical circumstances under which the criteria of usefulness have entered basic research and the subsequent developments that have created a market for basic scientific knowledge. We return to our empirical material to examine who has offered or "sold" science on this market, and how. Next we investigate how changes in international economic competition and its public rhetoric have influenced the balance between cooperation and competition, both within national research programs and between countries. Following that, we return to one of the persistent themes of the study, the unexpectedness of the discovery of HTS and the capacity of institutions and individuals to respond. We re-examine the topic of preparedness and the links between scientific creativity and the organization of research. Finally, we summarize the implications of these links for managing complexity.

6.1 How the worm got into the apple: basic science becomes useful

At present, the hegemonial quest for technological innovation has reached unprecedented scope in the minds of those who practice, administer, or apply science. Innovation has become a leading slogan in official rhetoric and informal discourse. Official documents in the United States speak of the erosion of US technological dominance and the EU justifies spending 3.4% of the total R&D budget of its member states with the need to enhance European industrial competitiveness. Japan is envied and attacked for its highly visible success in harnessing basic science to its high-tech industry.

Although it is generally acknowledged that many innovations take place "downstream" in the industrial process, by gradual improvements rather than spectacular breakthroughs, modern industrial societies have overwhelmingly opted for what has been the original contribution of science: its radical preference for the new. Ever new sources of knowledge are to be generated, tapped, appropriated, and exploited in the form of technological and social innovations. Basic science has become too valuable to be left to itself. It is seen as an essential ingredient, but only one ingredient, in the multi-faceted endeavor to generate technological innovations. New knowledge, insights, and instrumentation produced by basic science are no longer considered ends in themselves – if they ever were.

The relationships between science, technology, and economic growth have long been debated. Derek de Solla Price vigorously defended the claim that "the arrow of causality historically is largely from the technology to the science", rather than the reverse, as the conventional linear model has it (de Solla Price, 1983). Many historians of science and technology have made similar observations. More recently, however, the ground for argumentation in policy discourse has shifted. Basic science is being incorporated, possibly irreversibly, into a discourse on technology and economic growth. The preservation of a basic research base is considered essential to any modern economy. Skill-building occurs in this process, and scientific training is seen as a necessary precondition for attaining higher levels of technological activity (Pavitt & Patel, 1991). More recently, Nathan Rosenberg has turned the question around: Might it be a strong economy that leads a country to support its basic science system? Might it be that the willingness to invest in basic research is merely an aspect of economic policy? (Rosenberg, 1991: 345)

Such an interpretation, offered from the standpoint of the economics of research, is part of the increasing incorporation of basic science into the economic discourse of usefulness. We found abundant evidence of this shift in argumentation in the publicly and privately expressed motivations of practitioners of science as well as of science administrators. In our interviews with scientists during and immediately after the HTS euphoria, we were struck by the element of nostalgia in their excitement. It was as though they had fallen in love with science again after years of routine marriage. Their enthusiasm seemed youthful and innocent: everything seemed possible again, including the joy of pursuing science for its own sake – a far cry from the economic considerations that had come to prevail in their work.

Was there really ever a golden age of basic research, when science was free of crass economic considerations? The myth persists. Paul Veyne, in his classical essay, points out that the truth of a myth is always plural (Veyne, 1983). Every myth contains elements of historical truth, but much else besides. As research budgets shrink and the rapidly expanding number of researchers have increasing difficulties obtaining research grants, longing grows for a past imagined as free of such constraints.

The origin of the myth of a golden age in basic research can be dated and located in a period that was exceptional by a number of standards. US and European immigrant scientists in the Manhattan Project had contributed convincingly and terribly to the Allied victory in World War II. When peace came, they wanted to return as quickly as possible to their university campuses and laboratories, but also to retain the privileges they had enjoyed during the war, when their research had been generously and unbureaucratically funded. Science funding, it was argued, should be carried on to benefit the nation in other ways. Scientists saw no need for an administration to set priorities or to allocate funds. Except for defense research, the involvement of the military, who had occasionally functioned as adversaries and as de facto administrators, could now be dispensed with (Kevles, 1978). Scientists believed they could manage their own research, set their own goals, and allocate funds among themselves. All that was expected from the government was generous and continuous support.

Science – the endless frontier

This was, in essence, the content of the famous unwritten "contract" between US society and scientists after the war. Despite (or perhaps because of) its many ideological elements, it was a long-lasting consensus and produced effective results. The United States exercised political and economic dominance in the world, and for decades its technological base also benefited enormously from its scientific dominance. In response to President Roosevelt's request for a plan for postwar science, Vannevar Bush delivered his Report to the President in July 1945. Its title invoked an appropriately powerful American metaphor: *Science – The Endless Frontier* (Bush, 1945). The Report was a propaganda document, describing the enormous benefits that would accrue if government would continue to support science. In the interwar years, government had shown little interest in supporting science. Now scientists projected a world in which science would be just as important in peacetime as it had been during the war, opening up "that new frontier [that] shall be made accessible for development by all American citizens" (Bush, quoted in Shapley & Roy, 1985). The Federal Government would provide generous, stable funding especially for basic research in universities, to be administered by a board of university and industry scientists. The objective was to support military, medical, and physical science research. The National Science Foundation was established in 1950 as a result (Shapley & Roy, 1985).

Today, the Bush Report's definition of basic science provides an interesting point of reference for later interpretations. In a famous passage, Bush declared that:

> ... basic research is performed without thought of practical ends. It results in general knowledge and understanding of nature and its laws. This general knowledge provides the means of answering a large number of important practical problems, though it may not give a complete, specific answer to any one of them. The scientist doing basic research may not be at all interested in the practical applications of his work, yet the further progress of industrial development would eventually stagnate if basic scientific research were long neglected
>
> *(Bush, 1945)*.

For Bush, scientists working on a fundamental problem need not worry about applications, but "the person who funds the scientist" should.

The statement that science is "performed without thought of practical ends" can be read as a defense in the intense debate that followed Oppenheimer's remark that, in developing nuclear weapons, physicists had "known sin". It can be read as a vindication of the professional code prevalent among most scientists and engineers, few of whom were concerned about their role in society or the ends toward which their discoveries were used. The "technological optimists" (as Herbert York termed his mentor Ernest Lawrence) believe that "the sole business of scientists [is] to produce new knowledge and technology" (York, 1987: 75) and that anything else is an unprofessional waste of time. They think only pessimists harbor doubts about the obvious benefits of technology or compromise their professional commitment by worrying about the uses made of what they discover or produce (Salomon, 1990: 24).

A prevalent concern after the war was the link between basic research and technological applications. Shapley and Roy maintain that, in the context of US science policy in the 1950s and 1960s, basic research was interpreted as that "which has no foreseeable use", which would include much of today's research in high-energy physics, astronomy, and some branches of mathematics. They argue that this would be better termed "undirected" basic research, while research the sponsor expects to lead to applications, however far in the future, should be called "purposive" basic research. But somewhere along the historical trajectory, the distinction must have gotten lost. At some point, "the person who funds the scientist" started to worry continuously about applications and scientists could no longer take for granted that they would receive whatever funding they requested. They began to worry about their funders' worries. Basic research gradually became more tightly interlocked into the technoscience system as we know it today.

The expansion of science, organized innovation, and social accountability

The public mood has changed decisively in recent years. In the "advancement of science and its burdens", the burdens have come to weigh more heavily (Holton, 1986). Public disenchantment with science has grown, as science has become conflated with technology and its unwanted side-effects and risks. The pursuit and acquisition of new knowledge is no longer regarded as sufficiently valuable in itself to justify pouring money into basic research.

The terms of the unspoken contract between science and society are up for renegotiation – a process which has become more and more difficult. Pressure is mounting to anticipate both the potential negative impact of technology and the reaction of the public. Science is now subject to demands for public accountability for its future directions, its public acceptance, and its costs. Politicians and the public now want to know specifically what they will get for their money. Science is expected to prove its short-term economic utility. Only a few domains, like high-energy physics, are still perceived "without thought of its practical ends". Although scientists still cling to their own ethos of professional and scientific competence, rather than that of directly providing solutions for societies' problems, they have accepted the economic terms of the renegotiation.

Something has happened since the golden age of science, if it ever existed. Somehow the worm got into the beautiful apple of basic science, whose practitioners did not need to think about utility or responsibility. How did basic science, in the center of policy considerations after the war, recede into the background, while technological innovation has taken over almost the entire stage?

Bush's 1945 Report does not focus on the industrial uses of basic research, apparently because he assumed industry would seize on the findings of university and government laboratories. Products and applications would come as the "fallout" and "spin-offs" of basic science's discoveries. Ideas and opportunities would be the apples falling abundantly and continuously from the tree of knowledge, and industry would only need to collect them and watch over their maturation in the production process. At the time, when US science and technology flourished unrivaled, new science-based industries were emerging, such as chemicals, plastics, pharmaceuticals, and electronics; the age of high-tech firms was just around the corner. Wartime experience reinforced the linear model, i.e. it was widely believed that unexpected breakthroughs would occur in basic science, and that applied scientists and engineers could, more or less automatically, turn these into a continuous stream of innovative new products and processes. All that was demanded of science policy was the unfailing nurturing of this tree of knowledge, so that it would continue to yield its useful fruits.

And indeed, a period of unprecedented expansion followed. Organized innovation, a rare exception a century ago, became the rule. Postwar private and federal R&D spending soared, especially in the US.[1] With it, the agencies administering the science budgets expanded by an order of magnitude. The expanded role of government coincided with the emergence of "big science", exemplified in Europe by the establishment of CERN. In the universities, the pool of scientific competence, physical equipment, and facilities essential to high-quality research was vastly enlarged. This tendency peaked in the late 1960s and has since declined, sometimes drastically, especially in the United States.

Private industrial research in the United States retained its dominance and the number of scientists and engineers continued to grow. During the early postwar period, buoyant domestic and international markets supported robust profits and rapid expansion for large US corporations. Fundamental research was carried out in centralized facilities. Mowery and Rosenberg estimate that the share of US basic research performed within corporate labs may have peaked in the 1950s and early 1960s (Mowery & Rosenberg, 1989: 157). In a number of respects, the structure of large-scale R&D during this period rested upon a rigorous application of the linear model. Innovations were to be developed largely from internal sources within the firm. But the results were mixed. Firms that were not sufficiently insulated from foreign or domestic competition and that failed to cultivate interaction among their research labs came under increasing pressure. Industry's role as a site for basic research began to decline (Mowery & Rosenberg, 1989: 158).

Meanwhile, Europe and a newcomer, Japan, were catching up after a time lag. The growth rates of research council science budgets and of overall government spending on basic research were impressive. In the 1960s, European universities expanded to the point of explosion. In 1963, the OECD's Svennilson Report, *Science, Economic Growth and Government Policy* high- lighted, at national and international levels, the various presumed and measurable links between science and technology, economic growth, and the role of policy interventions. In 1971, it was followed by the influential Brooks

---

[1] As late as 1969, the combined R&D outlays of West Germany, France, the United Kingdom, and Japan were $ 11.3 billion, while those for the United States alone were $ 25.6 billion. The combined spending of these four countries did not exceed that of the US until 1979, when the figures were $ 58.3 billion and $55 billion, respectively (Mowery & Rosenberg, 1989: 125).

Report, *Science, Growth and Society*. The difficulty of demonstrating a clear correlation between research spending and the state of a country's economy became apparent. The assumptions of the linear model began to seem naive and out of touch with the complexities of reality. Attention shifted toward detailed questions about the role of technological innovation, government policy, and links between university-based science and industry. When some large-scale civilian projects, structured like military ones, confronted soaring costs, demand increased for the evaluation of research programs and for closer attention to estimated and actual expenditures. In 1963, de Solla Price's book *Little Science, Big Science* argued convincingly that the exponential growth of scientific activities could not continue indefinitely.

In part, it was science's expansion and growth itself that invited the worm to enter the apple. Budgetary requirements and the number of researchers had expanded apace. Funding agencies, whose numbers and bureaucratic power had likewise expanded, began expressing heightened expectations of returns from basic research. Moreover, this expansion had began to spread from the handful of industrialized countries who had monopolized scientific output after the war to the rest of the globe. The optimists had not considered that other countries might be able to turn out marketable products, too. So competition expanded from within national research systems to between nations. Bush's 1945 vision of an "endless frontier" for the United States rested on the simultaneously universalistic and ethnocentric assumption that the US was the dominant and virtually sole player in the world. In reality, the frontier could not and did not end at the Pacific coast.

Meanwhile, competition for research funds intensified as the number of researchers grew. Funding agencies began trying to predict the likelihood of successful projects and sought other ways to improve the efficiency and effectiveness of a research system in principle insatiable. Research needs and the demand for funding are open-ended. Michael Polanyi could still advocate that scientists should distribute themselves over the whole field of possible discoveries, but at some point the number of opportunities presenting themselves exceeds the scope of any funding system. The necessity had arisen to select, set priorities, judge economic value, and choose among a spectrum of options. The worm, once entered, proceeded relentlessly.

Markets and the marketability of science

Expansion is not restricted to numerical expansion within the science system, as de Solla Price fully appreciated. The real growth phenomenon is that of science expanding beyond its own boundaries. Science discovered markets, was absorbed and appropriated into markets, and created markets of its own. The latter include industrial markets for high-tech products, markets for students, and market niches for specialized knowledge. Science's unprecedented postwar growth is based in the increasing marketability, availability, and versatile utility of scientific knowledge and in the increasing demand for both general and highly specialized knowledge. It is now recognized that technological innovation and the diffusion of knowledge are based upon a close and intricate flux of ideas, methods, skills, instrumentation, and people, all of which are embedded in an entrepreneurial culture of production. The marketability of science came to be measured. Based on survey data from US companies in seven industries, Edwin Mansfield estimated that investment in academic research yields an average annual return of about 28%, and that the typical time lag from academic research to commercialization is seven years. He also shows that the research considered most valuable to industry does not always originate in the top academic institutions. Rather, geographical proximity seems to be a factor; effective communication between universities and industry is of prime importance in the successful transfer of knowledge (Mansfield, 1991).

The increased marketability of scientific knowledge manifests itself not only in quantitative attempts to evaluate the return on investment in basic research or to measure the social returns of academic research, such as the creation of jobs. Finding market niches is now part of the survival strategies of scientists who once believed themselves sheltered from such considerations. In recent years, the university has been turning into an academic market place. This blurring of boundaries has consequences for the culture of research. Mutual expectations and interdependencies tend to mingle norms. Especially in the United States, the boardroom has entered academia, at least in the higher, decision-making echelons, while collegial, egalitarian norms and outlooks have spread from academia to industry. A number of scientists have openly opted to become entrepreneurs. For those remaining in academia, possessing entrepreneurial talents and management capabilities is definitely an asset.

Quality control: selection criteria widen

A greater number of more heterogeneous knowledge producers now work at a greater number of more heterogeneous sites of knowledge production. This essentially sets the stage for one of the most profound changes the research system is now undergoing (Gibbons *et al.*, 1994). It also opens the door for a wider set of selection criteria in the evaluation of requests for research funding and prepares the way for basic science to move into heterogeneous contexts of application. As long as basic science was self-administered, scientific quality was the overriding criteria. With further expansion of the system, scientific quality became more difficult to define as the circle of evaluators also expanded beyond the guild of (eminent) scientists. Attempts were made to define quality, and additional criteria began to seep into the selection and evaluation process.

HTS: selling basic science

Even as this book is being written, HTS research is still performed as what usually passes for basic science. If we accept Michael Posner's definition of basic science as "that subset of what counts as good science which cannot be sold for immediate use", our study unveils a paradox, since all our evidence points in the opposite direction. HTS was sold to the public – more specifically to the taxpayer who foots the research bills – by politicians who themselves had been sold the prospects of proudly riding the wave of the next key technology. It was sold to and by the media as bearing the potential for almost limitless new technological applications – levitated trains, energy-saving power transmission systems, and entirely new gadgets. Dressed up in the garb of its future technological benefits, HTS had the advantage of convincing any lay person why, in this case, funds for basic research were needed and deserved. Other participants colluding in buying and selling technological promises were the research councils and government ministries responsible for funding research they consider vital to the national interest. But who sold what to whom?

In the case of HTS, scientists were not the only salesmen. Nor were they the magicians the media sometimes portrayed, nor the powerful lobbyists that are associated with "big science" or that emerge in the course of prolonged mutual service with governments. Of course, researchers tried to persuade funding agencies, politicians, and the media, and in this sense they lobbied. The "old" LTS specialists were initially highly visible in a field where no one could otherwise identify the experts. But their lobbying was amateurish and ad hoc, fueled, at least in the period immediately after the Zurich discovery, by their own naive "gold fever". Science had unexpectedly regained its excitement, part of which stemmed from the feeling that everything was possible; and hence everything could be tried out with a renewed sense of innocent irresponsibility. Although they publicly expressed confidence that HTS research would yield technological fruit, the researchers' mood was aloof, exhilarating in a sense of freedom from practical considerations. It was a period exempt from the deliberate planning that would all too soon become part of a researchers' necessary survival strategy again.

No one can be singled-out as originally "selling" HTS basic research for its applications, but everyone took part. Michel Foucault might have termed what took place a "truth game", in which each participant has his or her version of an elusive but shared belief. It was obvious to all who took part in the game that what they hoped to engage in, fund, or set up and manage would remain in the realm of basic science for years to come. No serious researcher or science administrator, no journalist adhering to professional standards could or did claim otherwise. And yet every public representation of the new research field, every speech or interview, even every international conference resonated with the hopes and expectations that technological benefits would come, sooner or later. The time horizons expected for the different types of technologies were discussed publicly and privately, and various estimates were published. This truth game was surprisingly high in truth content, but it remained a game, a collective bet on an uncertain outcome. What gave rise to the fervent, pervasively shared enthusiasm and belief? There was no proof to substantiate the belief, and no one claimed that it could be substantiated.

In an earlier chapter, we noted the force of the memory of the history of conventional superconductivity in a period of general technological optimism. But memories never suffice to chart a course for the future. Something else has to come into play. We believe that HTS highlights the extent to which new criteria were effectively used in judging which areas of basic science should be funded. The criterion of "good science" or "scientific excellence" was no longer enough to mobilize additional funds to launch research in a new field. Moreover, there was no one qualified and no criteria to evaluate scientific excellence in a field so new. In the countries we studied, the ritual of peer review was performed, but it met with a number of real difficulties in practice. Funding agencies and the political establishment wanted other criteria to help decide among the various research options and proposals. The increased political demand for science's accountability and for enhancing the country's international economic performance could hardly have been suspended for HTS research. These additional selection criteria were reinforced by international competitiveness for future high-tech markets.

Basic science can be appropriated by the political process only because scientists themselves have agreed to also consider the practical utility of their work. As we saw, many of them believed that Müller and Bednorz had not only made a beautiful discovery, but also that the Zurich oxides would eventually lead to technological applications. This indicates that researchers in basic science have learned to anticipate and favor research directions with technological promise.

6.2 Globalization, competition, and cooperation

Constructing the belief in the usefulness of basic science is a collective endeavor effective only if sustained by robust institutional structures and mechanisms. This makes basic science dependent on public funding to a vastly increased degree. It has always been accepted that industry would support whatever basic science it considered necessary to its profits but, in 1945, Bush pleaded for generous, long-term government support for basic science, with the aim of enhancing the nation's prosperity and its citizens' well-being. Although patterns of spending differ significantly among the United States, Europe, and Japan, all now spend increasing sums on basic research on the premise that this will fuel technological innovation and economic benefits to the investing nation – an assumption observers increasingly question. Harvey Brooks has remarked that the R&D system, even where successful, does not guarantee that the resulting economic benefits accrue to the investor (Brooks, 1987). Since national governments or private multinational corporations set most of the priorities for R&D (Pavitt & Patel, 1991), in the international arena most research activity is more competitive than cooperative. The spillover to non-participants is greater than the investors plan or desire.

National prowess and globalization

As Brooks put it, governments and companies still behave as if being in the forefront of scientific discoveries and technological inventions and innovation were key national assets to be nurtured for both military and economic advantage. Governments all seem to identify the same set of key technologies as necessary to maintain and further their sovereignty: microelectronics, communication, biotechnology, advanced materials, aeronautics, and space. National rivalries figure even more prominently in government-sponsored "big science" projects, which have become major features of almost all national science policies (Brooks, 1987).

The assumption that R&D benefits accrue to the nation investing is held especially strongly for basic research. It does not lack a certain irony if arguments about the economic returns from basic science are publicly put forth by scientists, while it is the economists who remain sceptical. It is difficult to judge the intrinsic economic value of basic research, technology transfer is now rapid, and the technological abilities of companies outside the traditionally technological nations are growing. This leads to an internationalization of the sourcing of industrial R&D, i.e. of the scientific input needed by firms (Mowery & Rosenberg, 1989).

In the first half of the 20th century, the United Kingdom demonstrated that high-quality science cannot compensate for the absence of complementary managerial and engineering skills. More recently, Japan vindicates – at least for the time being in certain industries – the strategy of the "fast second" who exploits the hard-won results of other countries' R&D.

We simply do not know how much high-quality basic research a national economy requires in order to remain competitive in the global economy of the late twentieth century, in which technological and scientific knowledge move more and more rapidly across national boundaries.... With it goes the reduced ability of any one national economy to appropriate the economic returns from basic research within its boundaries

(Mowery & Rosenberg, 1989: 291–2)[2].

Similar doubts can be raised about development work. Imitators can reproduce the innovations made in another country. Thus the innovator's monopoly is temporary. And lead time is shrinking, due to the rapid circulation of scientific and technological information. Recent studies have isolated a number of factors contributing to the loss of technological leadership, especially by the United States, and to the structural changes occurring in national R&D systems. They range from the expertise and skills base in companies, to labor relations, financial markets, and investment strategies (Dertouzos *et al.*, 1989). They include differences in managerial cultures and the roles played by governments in sheltering high-tech industries (Derian, 1990) as well as what constitutes comparative advantage in international trade (Krugman, 1991). The list could be extended but the result would hardly change: basic science and the economically rationalized policies supporting it appear almost incidental to a country's economic competitiveness.

International competitiveness and the belief in national investment returns

Why haven't these insights affected the widespread belief that R&D investment will benefit mainly the investing nation? Does the science system fear further cuts in public spending on basic research? Do politicians still seek the prestige of national prowess in science? The nation-state's tight embrace of science climaxed in the period between 1880 and 1914 and continued until the 1970s

[2] David *et al.*, (1988: 68–9) have analyzed the economic payoff from particle physics research, observing that:
    The outputs of basic research rarely possess intrinsic economic value. Instead, they are critically important inputs to other investment processes that yield further research findings, and sometimes yield innovations.... Policies that focus exclusively on the support of basic research with an eye to its economic payoffs will be ineffective unless they are also concerned with these complementary factors (*quoted in* Pavitt, 1991).

(Crawford et al., 1992) – does it linger on? National rivalries quickly flared up in the excitement about HTS and its technological promise, peaking in July 1987 when US President Reagan's conference on the commercial applications of superconductivity excluded foreign researchers. Though less drastic now, the atmosphere of rivalry remains, especially between the United States and Japan.

How can the discrepancy between an apparent economic reality and the incantation of a contrary belief be explained? Official policy discourse often aims to protect against uncertainties. The acceleration of technological innovation and rising numbers of international technological competitors has led to increased demands that domestic research institutes supply knowledge useful in staying ahead or catching up. This was clearly a major factor in the private and public HTS research initiatives. Visions of future technological applications speak a clear language. Joining the basic research effort from the beginning was equated with joining the next round in the race for technological innovations. This helped blur the line between basic research and applications and led to a merger of basic scientific and technological research agendas. At least some funders of HTS basic research would have preferred to fund technological applications from the start. Some have argued that only if it is embedded in a technological agenda can basic research lead to unpredicted new uses and devices.

Many of the researchers we interviewed repeatedly expressed the opinion that the Japanese research environment is better geared to merge and to exploit the merger of research agendas than is Europe, which they saw as still profoundly oriented toward basic science (and its beauty) for its own sake. Thus the reason why the international spillover of national basic science investment may be ignored may be that the merged research agenda is already a reality. The interchangeability in funding we observed in some European countries underlines this. How long will basic science be maintained as a separate research category at all?

As basic science and technological applications have moved closer, a mixture of technoscientific practices has emerged which is, by its nature, more subject to the political economy of technology. Thus basic science has come at least partially under the sway of rules governing technology. The economic and social forces influencing the development of large-scale technological systems and the convergence of research fields that have been characteristic of materials science are now impinging on basic research as well. Until now, the rule has been that "science knows no boundaries, but technologies do", meaning that technologies are subject to patent and property rights, legal jurisdictions, trade policies, and the right to secrecy – all of which threaten the free flow of scientific information across or even within national borders (Keynan, 1991). In HTS, this has not yet become an issue, but it is easy to imagine circumstances under which it would.

Over the last decade, the world market shares in some industries have shifted greatly.[3] While investment in basic science is an indispensable factor in fueling innovation, its returns are highly contingent, long term, and incidental to national economic prowess. No mechanisms exist to ensure that a country's investment will be reflected in its competitive position. But scientists can hardly be blamed for using this argument in public discourse, since other participants in the game also do so. Despite considerable evidence to the contrary, this argument still serves to legitimize national R&D funding. To give it up would be to invite the dismantling of the system – something the players can hardly be expected to do. While science is denationalizing (Crawford *et al.*, 1992), the science system remains overwhelmingly national in its orientation and structure.

Managing the external frontier

The argument of international competitiveness is thus central to the maintenance of funding for basic research. This thus demands the management of the external frontier – a country's scientific–technological research agenda and its relationship to potential foreign competitors who could appropriate its research results. What information flows where and to what effect? What technologies should be regulated by trade policies – and how? What knowledge should be patented, i.e. granted by the state as a temporary monopoly to an industry, and what should merely remain a trade secret, not protected by law, but also not subject to time limits? (Keynan, 1991) Answers to these questions are no longer as clear as when "science" still appeared to be a universally accessible "free good", while technology was seen as property. Special skills and in-house knowledge are often necessary for access, but these too are spreading rapidly. The nation-state that funds a national R&D system and wishes to see its own industry reap the technological benefits worldwide is faced with formidable problems, particularly when costs drive industrial firms to cooperate across national borders. Research policies for the EU must be seen in the light of these issues.

---

[3] In the early 1970s, US companies produced 95% of the telephones and 80% of the television sets purchased for US homes. In 1988, their share had fallen to 25% and 10%, respectively (Abelson, 1988). The US semiconductor industry has also eroded. In 1970, Japan had no share in the market for dynamic random access memories (D-RAMs), a semiconductor device, but by 1988 it had 80% (National Advisory Committee on Semiconductors, 1989). Although US firms continue to compete effectively in such fields as chemicals, pharmaceuticals, aircraft engines, etc., the trends worry US policy-makers and funding agencies. In Europe, all research funded by the EU is subject to the Commission's mandate "to enhance European industrial competitiveness."

New patterns of intensified international competition and cooperation between nations are arising. National governments cooperate, even if tendencies persist to control the utilization of research for national purposes. Governments want their researchers taking part in international cooperative projects wherever such cooperation is expected to yield tangible benefits for the country concerned. They are also willing to pay a kind of insurance premium allowing the country to take part in research efforts that may harbor strategic knowledge for future applications. This option is particularly attractive for small countries who want to keep watch on windows of opportunity that might open up.

European countries' desire to cooperate is partly rooted in an awareness of the constraints imposed by their small sizes. The smaller an industrial country's scientific community and resources, the greater its willingness to cooperate. Good ideas can arise in any country with a modern education system and an infrastructure for science. But ideas are not sufficient. This holds for university research if no industrial capacity is currently available as well as for countries lacking adequate industrial infrastructure. International cooperation is one way of overcoming such constraints. Scientists, technologists, and industrial entrepreneurs are eager to expand their horizons by "going international". They want to cooperate with the field's leading scientific teams. And firms desiring access to markets seek risk-sharing partners and foreign capital or technology.

Cooperation and competition in HTS

Researchers have always known how to balance their commitments to international science and to their own countries (or firms). Almost 100 years ago, Pasteur, a very patriotic scientist, wrote: "*le savant a une patrie, mais la science n'en a pas.*" Apart from reminding us that present day issue of proprietary rights and competitiveness had their equivalent in national hegemony and prestige, what has changed since Pasteur's days? Governments have traditionally agreed to leave basic science as an international playground for scientists; but this is now in danger of erosion through encapsulation or proprietary attitudes in research fields where applications are closely linked to basic science.

In the various phases of HTS research, researchers themselves have had no difficulty assuming international and national commitments and behavior at the same time. But they soon encountered familiar constraints. To obtain additional funds quickly, they had to lobby their respective decision-makers for national research programs. The more organized research efforts relapsed into the national institutional mold. Researchers parlayed internationally acquired information into "knowledge capital" in negotiating for funds within their own countries. When national plans were established, researchers' orientation necessarily narrowed to the local and national.

We expected that the larger number of researchers and scarcer resources would increase competition. And many of the strains in the current system are described in these terms. Funding agencies' increasingly short-term expectations for tangible results exert pressure to bring results. Leon Lederman's report to the Board of Directors of the American Association for the Advancement of Science (AAAS), appropriately titled *Science – The End of the Frontier*, points out that today's researchers are expected to publish more papers, take on more graduate students, and bring in more government funding than were their predecessors (Lederman, 1991).

Is this the case mainly for "routine science" and "routine funding"? Part of the excitement HTS generated among researchers was because a promising new field would inevitably entail new sources of funding and a reshuffling of existing funding patterns. At least briefly, the endless frontier reappeared on the horizon. On this basis we expected competition to lessen, at least after the hurdle of getting in has been passed. We expected that national research efforts would attempt to set up coherent programs that reduced competition within a country by channeling it outward to the international arena. What we in fact observed was that the level of additional funding made available for the new field remained much lower than expected. Although researchers invoked the sums provided in other countries to convince politicians and the public of the urgency of a national effort, funding agencies behaved as if they were an international cartel with an unspoken agreement on the upper limits of funding. On the other hand, the limited actual availability of university researchers raises the question of whether additional funds could have been used to good effect. In Europe, most university researchers chose to go into HTS research only on a part-time basis.

Citation counts show that, for reasons linked to overall funding patterns, US university researchers responded much more promptly to the new research field than either European university researchers or their own countrymen working in federal labs (Raz, 1987). The latter in particular were engaged in other research programs with long-term funding. As we have seen, industry everywhere proceeded cautiously. Exact figures are difficult to obtain, but it appears that industrial HTS research involved comparatively few researchers. In Germany, about 15% of all HTS researchers were in industry. HTS research may be a new and lively field, but nowhere has it been realized on a scale that public and scientific expectations seemed to demand.

Nor was competition for funds within a country as intense as we expected. Perhaps research communities adjusted their expectations when they realized that funding agencies would not open their coffers as readily as hoped. In Austria, researchers were widely disappointed at the disparity between the meager funds alloted to each research group and the expected commitment to submit to coordination. They often regarded HTS funding primarily as an opportunity to acquire badly needed lab equipment. Swiss university researchers, by contrast, were satisfied with the sums they received, especially given the infrastructure and instrumentation already in place.

In the first phase of the national research programs, the allocation distribution of funds varied in the three countries we studied. In Germany, the VDI functioned as the officially appointed intermediary, preselecting and selecting research proposals (which, of course, included referees). Switzerland, a small country with a relatively wealthy research system, was able to provide reasonable amounts of funding to almost everyone who wanted to be included, in the first phase. Austria, just as small as Switzerland but less well equipped in its university labs, used the shotgun approach, disgruntling the research groups. In the Netherlands, the first phase was relatively open to all who wished to take part. But it soon became clear that the second phase would impose much more stringent criteria. In the United Kingdom, published reports imply that HTS was seen as a test case for setting up new interdisciplinary research centers as instruments of government policy. Unsurprisingly, criticism of the new policy mingled with criticism of its implementation in funding HTS research. In the United States, funding and the competition for funds was also seen in the light of larger policy issues. Relatively early, the ongoing debate on "big science" projects surfaced; the "megalomania" of the enormous sums required for the SSC, allegedly cherished by politicians for publicity reasons, was contrasted with the paucity of funds available to "little science" (Anderson, 1991). But the debate about adequate funding and competition was overshadowed by the technological and economic rivalry between the US and Japan.

Channeling competition outward

Creating islands of national cooperation in a sea of international competition amounts to avoiding excessive competition within a country by channeling it outward. Although in the early phase after the discovery of HTS individual research groups stepped out of their nation's shadow to engage in intense communication, exchange, and competition, they were soon "recaptured" to compete in the more coherent form of a national program.

Two mechanisms were decisive in lowering the level of internal competition and channeling it outward. One was the surprising flexibility with which university research groups turn toward any new field of research. In HTS, entry requirements were low. All groups with LTS experience could claim access – but so could almost anyone else with a background in physics, chemistry, crystallography, or materials science. The largely exploratory nature of the first phase of the national research programs meant that research proposals had to be accepted more or less at face value. Without clear quality standards or indications of which directions were most promising, competition was necessarily diffuse. Funding was broad and reputations were rapidly gained – and rapidly lost. Reputations proved stable only in the top echelons of international science, where cognitive and not social hierarchies matter.

The second mechanism was inherent in the funding conditions laid down for the second phase of the national research programs. These resulted from the activity of intermediary bodies and from the typical configurative environment in which hybrid communities thrive (Gibbons *et al.*, 1994). The overriding selection criterion for basic research was the promise of technological application – however difficult this was to prove, except in intention. Links with industry were reinforced, giving the programs stronger goal orientations. The national "teams" dramatized by the media soon began to differentiate: the amateurs dropped out, and those remaining became professional players.

The first phase had allowed the exploration of a new territory whose exact nature, fertility, and extent were unknown but extremely promising. The new trails are evident in the "hot" publications, the papers cited most often (Pendlebury, 1988a, b). Gradually the territory became more familiar and more densely populated. In an investigation of the differences between scientific disciplines, Tony Becher noted great variations in the density of populations (i.e. researchers) in given specializations and the related patterns of how competition is expressed (Becher, 1989). In some specializations, few issues are pursued, even if the number of researchers involved may be high. In other areas, there may be a virtually unlimited number of questions that may reasonably be expected to be answered, and yet a small number of people pursuing them. Becher called these variations the people-to-problem ratio, and used the metaphor of urban and rural life styles in research. In urban problem areas, a densely concentrated population of researchers engages in intense, concentrated, collective activity and lives at a hectic pace. There is severe competition for scarce problem space and resources, but also a rapid, heavily used, and efficient information network. Urban-style research is characterized by intense competition *and* teamwork. Where it is easy to avoid overlap, joint activity is less common (Becher, 1989).

In the case of HTS, the new cognitive territory was initially empty and open. But step by step, new trails and signposts began pointing toward potential technological applications. In the best-surveyed area, sensors and other instruments for processing and quality control in manufacturing seemed in the offing for research, health, and defense. SQUIDs for measuring magnetic fields have possible applications in surveying, surveillance health and research. A–D converters, computer wiring, and entirely new devices would find applications in the electronics and communication sectors. Many challenges remained in combining semiconductors and superconductors. This may be HTS' urban zone, where the search activities of many researchers come together. In an intermediate and less-settled zone, magnetic separation – already demonstrated in kaolin clay purification – could be used more widely in pollution abatement, or new applications could be sought in transportation, with superconducting elements offering advantages over conventional magnetic levitation systems. Superconducting devices and materials, suitably developed, might be incorporated into electric utilities.

But perhaps HTS is best described as a high-tech science park located on the outskirts of a city, with good communications facilities but plenty of space to expand. Teamwork is neither as competitive as in the dense, urban mode, since many options can be pursued, nor as loosely structured as in the rural mode, since demands for technical feasibility and commercial viability exert powerful constraints. In the best of all possible worlds, many such centers could thrive independently while remaining closely connected.

Creating the conditions for research in an international context

Scientists' conflict of loyalties between the internationalism of science and their commitment to contribute to the international research effort on behalf of their own country or company was therefore resolved in a familiar way. As Keynan has noted, in most democratic countries, scientific communities and their organizational frameworks run their own affairs independent of governments. In most of these countries, it is also accepted that the progress of science depends critically on the free flow of ideas and basic scientific information across political borders. Most politicians, political parties, and political philosophies may not fundamentally understand the notion of academic freedom or the universality of science, but they do realize that government interference in the conduct of science is counterproductive (Keynan, 1991).

For their own purposes, scientists reinforce and exploit the contradictions between government research strategies based on competitiveness and the more formalized international cooperation in areas that promise to serve the national interest. As active lobbyists in policy-making, scientists use whatever arguments they think might persuade the politicians. They act as consultants to industry and as advisors to governments. They take an active part in setting up and managing national research systems as well as international collaborations. Occasionally they astutely manipulate the rivalry between nations to obtain scientific facilities or programs. Scientists also act as highly skilled research managers in research councils or similar funding bodies, where their intimate knowledge of university, industry, and government systems makes them good intermediaries and networkers.

How does international science encounter national interests today? How does the increasing internationalization of R&D capacity affect international competitiveness and the management of the national–international boundary? Is there a better alternative to the existing national funding system, and if so, what would it be?

Since Pasteur's day, scientists have maintained undiminished allegiance to the internationalism of science. But their homeland, although still the primary source of needed research funds, no longer commands their loyalty as much as it once did. It is clearly being displaced by something else expressed on the business pages of the press, both national and international. Their country calls continuously for cooperation with industry, which again can be national or multinational, and prepares them for worldwide economic competition, in which science is only one of many ingredients for success. Today, basic research has moved much closer to or even merged with a technoscientific research agenda. Technology and the skills to exploit scientific–technical knowledge have spread and continue to spread from the formerly leading industrial nations. The extended lab has reached global dimensions, though different parts of this lab operate under very different conditions.

Paradoxically, economic competitiveness also enhances international scientific collaboration. Robert Reich has argued that the days of economic nationalism may be numbered and that we may be living in a period of transformation of the politics and economics of the coming century. National products, technologies, corporations, and industries would be a thing of the past. Each nation's primary task would be to cope with the centrifugal forces of a global economy straining the ties binding citizens (Reich, 1991). Scientists and other "symbolic analysts", as Reich terms them, would be among those best able to thrive in a world where national borders have no economic meaning, and they would be more tempted to discard the bonds of national allegiance. But if this is the course taken, science itself will be transformed. No longer at the service of economic nationalism, its inherent internationalism and universality would emerge in the service of economic internationalism.

6.3 Scientific creativity, preparedness, and the organization of research

Any discussion of whether the Müller and Bednorz discovery can be interpreted as "mere chance" or "programmed success" must start with carefully defined premises. Even then, we are still faced with the fact that the process of discovery is at the same time intensely individual and collective (Nowotny, 1990). It is the individual scientist who embarks on an untrodden path, but he or she uses all the available support – theories, techniques, instruments and equipment embodying scientific knowledge, as well as the more tangible resources of a lab and a position within it – that the collectivity of science can offer. In this section we focus on two questions: how is the discovery of HTS to be interpreted in the light of the inherent tension between individual scientific creativity and the collective nature of science? And can the social organization of research enhance scientific creativity?

Scientific creativity and the individual scientist

The predominant theme in the history of science and in the literature on scientific creativity is the individual scientist. As Steven Shapin has noted, to portray the individual scientist as working heroically alone or with an assistant fits the image of "three hundred years of solitude" in the allegedly public culture of science (Shapin, 1991). Many examples, anecdotal evidence, and biographical or autobiographical narratives document the difficulties of giving a public account of what scientists themselves see as an intensely private experience. The more meager the documentation of the circumstances of discoveries, the wider is the scope for differing interpretations and for the creation of myths. Simon Schaffer has recently drawn attention to the intriguing relationship between the authorship of discovery and its authorization as a discovery, in which retrospective judgment plays a crucial role (Schaffer, 1994). Although attention has been turning to the topics of scientific intuition, images, and themes that pervade a scientist's work and life, and although much more is known today about the microstructure of scientific creativity, the difficulty remains that we can speak about it only in hindsight (Root-Bernstein, 1989; Gardner, 1993; Boden, 1994). It is as though the mystique of creativity can only be preserved if it does not yield its last secrets.

Every biographer is faced with the distortions inherent in the reconstruction of private experience. Scientific biographers are subjected to the additional temptation to seize upon seemingly trivial details to make the hero–scientist seem more normal and approachable. Müller has not escaped these dramatic elaborations. Little knowledge is gained by hearing that "for most of his life

Müller has relied on his strong intuitions about nature, good feelings that he feels absolutely compelled to test", and that he has "been drawn to the works of his countryman, Carl Jung. Every three or four months he gets together with his brother, a psychologist, to discuss consciousness, the human spirit, and other such numinous concepts" (Schechter, 1989: 66). Müller is also quoted as saying, "If I ever stop being able to do useful science I would like to devote myself to these things full time" (Schechter, 1989: 70). Does it simply make Müller "more human" if Schechter also reports that, since his twenties, Müller has carefully compiled several thousand pages of his dreams? In his Nobel lecture, Müller unsurprisingly referred not to his dreams but to the Jahn–Teller polaron model as having been his "guiding concept", even if the latter proved not to have worked as they expected (Bednorz & Müller, 1987: 70).

Paul Chu's search for substitute compounds, his ingenious red herrings to keep referees and colleagues from learning his "secret" too soon, and his victory in the race to reach a critical temperature at the liquid-nitrogen level all made for good stories and turned him into a media star (Kolata, 1987). But what do the many incidents and coincidences of that short, intensely competitive race tell us about chance and programed success, except that the route is lined with personal ambitions, cunning strategies, petty jealousies, the ability to "sell" oneself to the media as a necessary concomitant of modern science relations, and that the world is unjust? The persistent public image of the lonely hero of science lives on in the young, aggressive Chu and the quiet, aging Müller patiently following his intuition.

HTS was not serendipitously discovered by researchers who were not looking for it. Müller and Bednorz were conducting systematic research, though as relative outsiders and against mainstream opinion. Theirs was not even a case of an unexpected outcome of a planned experiment, since Müller and Bednorz obtained precisely what they were looking for. What makes their discovery interesting from the perspective of social studies of science is its embeddedness in the interface between private experience, intuition, individual drive and the social organization of public scientific knowledge, i.e. the practice of science. The latter is more than the organization of a research lab. It pertains to a field and its traditions, its history of successful and failed concepts, of technological applications sought and sometimes found. After all, an expert is someone who has explored many avenues and knows why they don't work, and who incorporates the public knowledge of errors and successes into his or her own private knowledge.

On the private side of the equation, Pasteur's dictum that "in the field of observation, chance favors only the prepared mind", still holds. For him, the "prepared mind" was the mind guided by theory: "Without a theory, practice is only routine driven by habit. Theory alone can cause the spirit of invention to appear and develop" (Root-Bernstein, 1989: 87). But Pasteur also knew that a theory or "guiding concept", as Müller and Bednorz called it, can only show the direction in which to guess. "The illusions of the experimenter form the greater part of his power. These are the preconceived ideas which serve to guide him" (Root-Bernstein, 1989: 139).

Interaction between private and public scientific knowledge

But preparedness also has a public side. First, there is the interaction between private and public scientific knowledge. We may never be able to tell why, possessing the same data, one individual makes a discovery and another doesn't, but we can attempt to reconstruct the social and individual aspects of what Jürgen Renn subsumes under the term of cognitive structures. Basing his analysis on the works of Galileo and Einstein, Renn maintains that cognitive structures are not universal, but historical, and depend on the given material tools of representing knowledge, such as writing, printing, or other information-processing technologies. Without these material representations, knowledge cannot be communicated between individuals, nor transmitted in history. Cognitive structures result when an individual appropriates the knowledge available in a given historical period. Their variants are the result of modifications and extensions by individual experience. The concrete examination of examples that were part of the communal scientific enterprise of a given historical time – including the socially available knowledge, its material means of representation, and its appropriation by a given individual – allows a reconstruction of cognitive structures.

Renn has shown how Galileo took part in the integration of the technological knowledge of his time, represented in the artisan tradition and the philosophical and scientific knowledge embedded in the intellectual tradition. Such a reconstruction illustrates the relationship between the social accessibility of knowledge and individual intellectual progress. Individual cognitive structures frequently "happen to be suited" to provide sometimes epochal solutions to problems, though these structures emerged as individual variants of socially determined knowledge structures in the context of intellectual efforts to solve quite unrelated problems (Renn, 1989).

On more mundane levels, some striking parallels of "finding direction by indirection" emerge in areas seemingly far removed from the cognitive structures that influenced Galileo or Einstein, like firms engaged in technological innovation. Often, technological innovation involves solutions to problems typically "ill-structured", i.e. where available knowledge and methods do not point to any obvious solution. Solutions require a reconfiguration of previous experience and available scientific knowledge. The problem-solvers must have both specific and uncoded capabilities (Dosi, 1988: 1126).

Technologies and industrial sectors differ markedly in their degrees of openness and universality and in the tacitness and specificity of their knowledge bases. Companies produce and utilize innovations not by simply dipping into a general stock of publicly available technical knowledge, nor do they survey the whole universe of knowledge before acting. Rather they seek to improve and diversify their technology by exploring areas that enable them to build on their existing technological base, in-house knowledge, and knowledge of external market conditions. Successfully combining proprietary, company-specific, and publicly available technological knowledge to solve poorly defined problems seems to be the key to the innovation process. But there remains an irreducible element that cannot be accessed, bought, or sold at will: the cumulatively augmented abilities, skills, experience, and private, tacit knowledge of individuals.

In scientific discoveries, the available knowledge base is overwhelming. Sir Brian Pippard (Brian Josephson's thesis advisor) has recounted the succession of ideas and theories on superconductivity when Josephson was working on his breakthrough theory (Pippard, 1977). Timing – stumbling upon the ideas of other scientists at the right moment – was crucial to his success. But so was "the debt ... that good scientists owe to other people's bad ideas" (Pippard, 1977: 12). Even wrong ideas are sometimes stepping stones on the path out of an intellectual labyrinth. In the case of HTS, Müller and Bednorz drew upon the public knowledge of an article by French colleagues; it described precisely the material their "guiding concept" told them to look for. Finding this article was undoubtedly one of the "fortuitous moments" that privileges the "prepared mind". When Raveau later re-examined his sample, he found superconductivity. But what would have happened if the article had not been available precisely when Müller and Bednorz needed it? The discovery of HTS would have had to wait a little longer.

What conditions can reproduce such creativity? What are the social inducements or hindrances enabling or preventing such events? What can be done collectively to grow prepared minds? Policy-makers are keen to know, but not all insights and generalizations can be translated into policy. The process of translation is hindered by the inherent tension between the social organization of scientific research – which is set up as a stable framework for reasonable expectations and predictable, reproducible results – and the unexpected, unpredictable discovery. By definition, the innovative break-through escapes predictability. Since it cannot be foreseen, it cannot be controlled. It eludes administrative competence, though the administrators of science seek to facilitate its occurrence. Much of what is known about scientific creativity follows only the logic of hindsight, so the rate of new discoveries cannot be increased at will. Like much other knowledge in the social sciences, it cannot be directly utilized.

The unexpectedness of the discovery and the novice effect

We noted that the discovery of HTS was unexpected in terms of its discoverers, the site of the discovery, and the ideas involved. It contradicted the conventional wisdom and general expectations of scientific peers and science administrators, but it falls into a familiar pattern in the history of science. Müller and Bednorz were outsiders; Müller a specialist in another field. Many discoveries have been made by scientists just entering a field, either the young or older scientists shifting focus. This has been termed the "novice effect" (Root-Bernstein, 1989: 417).[4]

The novice is able to see a problem in a new light. The practical lessons of this are simple, but far-reaching: to encourage scientists to change fields every ten years or so, to foster other kinds of intellectual mobility in scientists, and to encourage the interdisciplinary migration of concepts and practices: "To the extent that prediction is impossible, mobility is necessary" (Jacques, 1990: 147). Science ages with people, and may become a kind of refuge for those who cling to old ideas, a process currently visible in many scientific institutions whose funding has been frozen or reduced while staff levels are maintained. Mobility here is above all the mobility of ideas and of scientists prepared to extend their areas of competence, to retain their curiosity, and to enter new fields. That this is also often tied to geographical mobility is borne out by Müller, who otherwise would have provided a model of an aging, sedentary researcher. Without his 20-month sabbatical in the United States – with all the intellectual freedom such a stay entailed – it is unlikely that he would ever have embarked on the path he chose.

Contrary to the widespread belief that young researchers hold a monopoly on scientific creativity, it is not chronological age *per se* that predisposes to discovery. Older scientists tend to be overwhelmingly engaged in administrative duties, peer review obligations, fund-raising, and other forms of participa-

---

[4] Pasteur's germ theory of disease and physicist Luis Alvarez's theory that meteorite impact led to mass extinctions are two examples. Alvarez wrote, "I will probably be remembered longest for the work done in a field about which I knew absolutely nothing until I was sixty-six years old. This field is geology." Alvarez believes his ignorance was crucial to his discovery, since the technique he used – measuring iridium contents – had already been tried and found unsatisfactory by previous investigators (Root-Bernstein, 1989: 417).

tion in science policy. Whether this is due to changing preferences or is unavoidable in professional advancement, they are more involved in what Krohn and Küppers (1989) have called "scientizing". Young researchers tend to be open in their perception of problems and as daring in their proposed solutions, while older scientists naturally tend to develop some closure of their ideas and approaches.

Terry Shinn's compelling study showed that cognitive and social hierarchies in a French lab. were related to the research results obtained. The cognitive hierarchy complemented and competed with the social hierarchy. Especially at the top level, the results produced, the style of presentation, and the audience addressed were all determined by the positions occupied within the social and cognitive hierarchies (Shinn, 1988).

Müller broke out of the social hierarchy, looking for a new place in the cognitive hierarchy. A scientist's position in a laboratory and his or her specific place in the social and cognitive hierarchies provide different degrees of freedom. Only after Müller was appointed as an IBM research fellow did he enjoy the autonomy to set his own research agenda, and it was his desire for more time to pursue what he called his "own work" that led him to resign as a manager. But freedom may also come from moving into a new field, even in old age. Again, the history of science is replete with examples of scientists who made major contributions after entering a field at an age above average.[5]

The unexpected site of the discovery of HTS is tied to social conditions that provide more creative environments. The IBM Rüschlikon lab, though part of a large-scale research enterprise, was relatively modestly equipped. In this respect, Müller and Bednorz had plenty of company. Many breakthrough discoveries have been made with inexpensive equipment (Root-Bernstein, 1989: 396). At least in "little science", some of the most important developments have been achieved in poorly equipped, inadequately funded labs with seemingly amateurish techniques. But this cannot justify cutting research funds, since exploratory research aims to show that a phenomenon exists; if successful, it is usually the starting point for another line of research that will need considerably greater funding to yield profitable applications. HTS provides a good example.

Müller and Bednorz confirm another lesson from the history of science, one with strategic relevance and concerning the autonomy the HTS discoverers enjoyed in their work. Over the years, governments and large corporations have learned the wisdom of funding not only research leading directly to

---

[5] So it is not the number of years of life, but the number of years in the field that determine a scientist's "youth" in this sense. What scientists who made important discoveries past middle age have in common is that they changed fields more or less regularly, enjoying the freedom of novices again, but with the ability to apply the life-long experience and skill gained in another field.

applications, but also basic research whose benefits are unforeseeable. To some degree, research managers have also learned that it may be wise to grant individual scientists a measure of autonomy. The decision as to how much autonomy will be granted is made in the tension between the need to prepare for the unpredictable and the need to plan scientific research. Discoveries are surprises; all that can be planned is work that *might* lead to discoveries. Autonomy for the individual scientist is a prerequisite in this kind of planning, since he or she must be free to follow their own perhaps idiosyncratic strategies of combining private and public knowledge, utilizing whatever appears promising from the collective pool of theories and experimental techniques. Sufficient time is an important factor in exploring and following hunches. Administrative duties, if carried out personally and not simply in a supervisory capacity, do not go well together with research. A researcher needs to be able to dip freely into the wide field of existing opportunities, rather than to follow the pathways preordained by a research program. Such autonomy requires institutional niches deliberately carved out by the scientists or their administrators (Nowotny, 1990). Organizing the preconditions for research includes more than fund raising.

Proliferating options and planning research

Many science administrators have expressed the belief that Müller and Bednorz would never have received funding if they had submitted their plans in a research proposal. Administrators want to know what kind of results can be reasonably expected in what time span. But with exploratory research like that undertaken by Müller and Bednorz, it is not possible to say precisely how the work will advance. In this respect, such research resembles much social science research. In such a case, it is likely that scientific understanding of a phenomenon will be deepened, but it is not possible to state honestly what the results will be. This poses a problem for peer review. If exploratory research follows different rules, its proponents and results should be evaluated differently. The general practice, especially in the US system, exposes a dilemma. Pre-performance review tends to be conservative, and post-performance evaluation is extremely difficult. *Ex post* evaluation generally involves measuring results against objectives. But exploratory research cannot specify its goals in advance, except in very general terms (Cozzens, 1990).

The labyrinth of science presents far too many paths to permit systematic exploration. This is why theories and guiding concepts, hunches and intuitions are so important, as in the case of Müller and Bednorz. But when progress is being made, it opens up new avenues and options.

The idea of planning success in science was still anathema in the 1930s, during the fierce ideological battles between J. D. Bernal, who stood for those who wanted to harness science to "social functions", and Michael Polanyi, who wished to safeguard the "freedom" of science against perceived losses of its privileges. Today, every government, corporation, and university president wants to "program science" with plans for financing, staffing levels, and research lines. It is still scientific research that poses the questions to be pursued, but other social and economic considerations are increasingly drawn into the phase of problem definition, thus shaping the objects of science, both as intellectual objects and technical artifacts.

By its nature, basic science must remain uncommitted and open to potential technological applications. The two are now more closely linked, and the time in passing from one to the other is shorter. As options proliferate, later steps have to be considered more carefully and spelled out in advance. With increases in total research funding, budgets have become more complex, research programs require more coordination, and people – the most important resource – have to be trained and motivated for scientific work that requires them to be more mobile, fungible, and equally disposed to many options. Scientific and technological development does not follow a rational plan, but neither is it controlled by sinister economic forces. It proceeds according to a built-in logic favoring those who can take advantage of the opportunities that arise through new theories, insights, experimental results, and their interconnectedness, as well as through opportunities shaped by planning and coordination. The opportunities are created not only by scientists, but also by decision-making bodies, parliaments, ministries, and corporate boardrooms. The case of HTS illustrates some of these connections.

Most countries have some kind of national research plan, federal initiatives, *Schwerpunkt* programs, or the like. Thus the unforeseen is the initiating impulse that sets the system in motion, again and again. The fine-tuning in funding agencies, labs, and industry is needed to promote the unforeseen. But it is the fate of the unforeseen discovery to be superseded. Selection criteria such as cost, performance, reliability, and the scale of application become more important.

As many reconstructions of processes of innovation and diffusion have shown, no technological development takes off before it finds a new niche or pushes its way into an old niche. The importance of chance decreases over time, or rather in proportion to the solidification occurring when technological applications come to market. While chance favors the individual, programs help the institutions.

6.4 The innovation machinery of science at work

Program managers or analysts eventually encounter the decisive snag of the unpredictability of scientific and technological knowledge. All they can do is to seek to create conditions under which creativity can flourish and favoring the occurrence of what cannot be forced. Even if this can be done on a large scale, the problem remains of how to turn highly individualistic bursts of scientific creativity into socially desired technoscientific outcomes.

Widening the basis for individual creativity

What are the ideal conditions to foster creativity? Policy discourse speaks of supporting designated "centers of excellence", encouraging more interdisciplinary research, or strengthening university–industry ties with the idea that novel solutions are most likely to emerge in confrontation with practical problems. The importance of research management and the right leadership style are increasingly recognized, and maximum mobility of researchers and their ideas is encouraged. All these ideas are based in general experience; there is as yet no hard proof of their efficacy. In fact, Müller's and Bednorz's working conditions hardly fit the conventional image of the ideal. Their group was the smallest possible, below anything that policy-makers believe is the critical mass. Their interdisciplinarity consisted in Müller finding an assistant with skills complementing his own. The IBM lab in Zurich, while of good repute, was never declared a "center of excellence" and, as a pre-retirement Fellow, Müller worked in relative isolation. Except for his decisive stay at the Yorktown Heights lab, his mobility had been unusually low. So the IBM lab at Rüschlikon, widely envied, imitated, and considered symbolic after producing two Nobel Prizes, does not provide a model for generalization.

What can be learned from the HTS discovery, despite its unpredictability? Müller and Bednorz persisted for more than four years toward a goal they set themselves, in the face of a discouragingly recalcitrant phenomenon. They defined their situation as one in which they could risk failure – a luxury for others, whose untenured positions force them to keep up a steady flow of publications. Müller and Bednorz could pace themselves. Isolated and untouched by the ordinary pressures of work and success as they were, Müller's scientific interest in a field new to him had been sparked by his invitation to the US lab, where colleagues worked on the very practical problems of superconductivity. Mobility emerges again as a decisive factor.

Why can't such conditions be created on a broader basis? Why not create more working situations resembling the Zurich model? Of course this has been tried. But all attempts to collectively organize the individualistic mode of producing creative scientific work meet with an inherent constraint: the risk of failure is real. The question is: Who will assume the risk? If the risk of failure is restricted to a few individuals whose idiosyncracies are tolerated by colluding managers as long as they fulfill their other duties, creativity can survive in unofficial niches in official institutions. If they are successful, the organization shares the glory. If not, failure remains inconspicuous, since the risk was never officially acknowledged. But if a research organization or funding agency officially decides to take risks on a larger scale, it is accountable. Institutions are charged with limiting the occurrence of failures, which means limiting risks.

The preparedness of institutions

What does HTS show about the collective research response to the unpredictable? How well-prepared are funding agencies, research councils, and similar bodies when the unpredictable breakthrough occurs? How well-prepared are research groups to exploit the new opportunities? How well-prepared are countries to redirect their organized research efforts?

The one generalization safe to make in answer to these questions is that preparedness depends on the strengths and weaknesses of the historically evolved structures in which research is conducted. This may involve the equipment needed to do lab work, or the flexibility with which research groups reorient themselves and their graduate students.

There is a basic level of essential preconditions for research effectiveness. As Franklin has noted for European research systems, few projects or groups fall below this floor. There is also a ceiling above which few projects or scientists rise. The gap between floor and ceiling, and thus the variation in individual effectiveness, increases with improvements in research circumstances. Thus, even the best researchers are ineffective if facilities are too poor, but not even the best facilities can guarantee good research results (Franklin, 1988: 318–19).

Another aspect of preparedness is a research system's flexibility, including the components' flexibility toward each other. Flexibility presupposes free time or a shift in commitments and the ability to break out of bureaucratic routines when necessary. Different attitudes toward risk-taking may facilitate or hinder an institution's preparedness to respond to new opportunities. Universities are often credited with a much higher degree of flexibility than industry, but preparedness can be limited by the load of teaching and administrative duties, as well as by unstable prospects for graduate students due to fluctuating funding. Industry is more richly supplied with qualified personnel and long-term funding, but company goals can preclude responding to new research opportunities.

A country's preparedness finally is connected to its willingness and ability to enter the next round of competition. Essential are a general openness toward and awareness of the practical problems that are the missing link between basic research and technological applications. Researchers must be open to the interests of industry, and industry must be willing to enter promising lines of research. The industrial response, or its perceived likelihood, has been an important precondition for additional funding of basic HTS research. But industry's response cannot be planned either. In some cases, industry has been the weak link in the overall national response.

The innovation machinery at work

HTS has revealed the close linkage between basic science and technology, and how the anticipation of technological advance affects the choice of what basic science will be pursued. The extended lab continues to expand – neither according to a master plan nor by being left to chance. It is increasingly recognized that creativity and innovation depend on a self-organizing capacity. In the search for technological innovation, specific products are no longer targeted; instead, infrastructures and research conditions are planned with the intent to allow certain types of products to emerge – including those whose uses are as yet unknown. This points to the importance of linking technological innovation to social innovation. Potential innovation sites are systematically multiplied in society. Materials science provides a good example of technoscience's sprawl into new contexts and directions. And HTS fits well in this image of the extended lab.

Throughout our study, we have maintained that HTS allows us to identify a number of features characteristic of the current restructuring of the science and research system. HTS is not unique in involving the identification of features not found in nature as such but produced by our manipulation of nature. As a research field, HTS overlaps with a wider field, that of materials science, where solid state physicists, chemists, electronic engineers and others seek to understand, manipulate, and control nature at the microscopic level. Taking cues from biology and the self-assembling capacity of molecular structures, and seeking to exploit the knowledge of quantum mechanics for practical ends, these scientific practitioners seek to create devices that operate on the atomic level or on a nanometer scale. Quantum structures known as quantum wells, quantum wires, and quantum dots are now in use or being developed in conventional semiconductors. The limits to development are set by the sizes of atoms. It has become possible to move atoms individually with the aid of a scanning tunneling microscope.

The ability to manipulate and control atoms and hence to manage complexity in this realm of nature finds a parallel challenge in the necessity to manage complexity in the organization of research. We have repeatedly emphasized the high degree of fluidity in the organization of research, its exposure to heterogeneous contexts, the expectations of returns, and the self-organizing capacity manifest in the shared beliefs that helped to set up a new research field. By bringing together the perspectives of the different actors who mobilized resources on different levels and with somewhat different intentions, we have also attempted to show the synergetic effects or convergence of the fundamental changes that science and the culture of research are presently undergoing.

The convergence of these various dimensions in the case of HTS – perhaps in a unique configuration – provides a detailed view of the innovation machinery of science at work. Hans-Jörg Rheinberger calls the experimental sciences "generators of future" in the sense that they acquire their meaning only from what will have been. In the life sciences, for instance, the significance of mutation can only be judged after time. This is an example of what François Jacob calls, "*le jeu des possibles*", the game of possibilities (Jacob, 1981). Rheinberger says:

It cannot be performed purposefully and selectively in advance. Either one plays it, or one doesn't play it. And if one plays it, it cannot be brought to a closure because of its peculiar structure. For in order to know what one has done, one has to have entered the next round. This....is the central mechanism of modern research, its mode of bringing new things into being

(Rheinberger, 1994: 15).

The success of setting up research innovation machineries is based upon one of the most remarkable features of science, its ability to produce and proceed by consensus. There is an unusual degree of stability in the sciences, produced as a result of what Ian Hacking has called the self-vindication of the laboratory sciences (Hacking, 1992). The theories of the laboratory sciences are not directly compared with "the world"; they persist because they are true to phenomena that have been produced or even created by apparatus in the laboratory and measured by instruments engineered to this purpose.

As an integral part of such a powerful innovation machinery, the case of HTS reveals what may be the most disturbing paradox: the relative decline in the importance of the individual creative act, be it the individual scientific discovery or a technological invention (Nowotny, 1995). The establishment and functioning of an innovation machinery that exploits the fruits of individual scientific creativity shows that the contexts of individual scientific creative production have multiplied and expanded in scale. It is certainly no longer a heroic age of the individual creator. The new heroes proliferate and interconnect. If individual scientific creativity is a necessary but no longer sufficient precondition in a long, branching sequence of possibilities – *le jeu des possibles* – then the machinery that has been installed usurps the role previously held by the individual discoverer or inventor. The machinery is set up not only to produce planned but unpredictable innovation, but also to exploit such innovation.

The enthusiasm with which the act of individual scientific creativity in the discovery of HTS was received is poignant in the face of the profound changes in the science system's operation. As indispensable as individual scientific creativity is and will remain, it has nevertheless become increasingly contingent on what the innovation machinery of science will make of it – after each new breakthrough.

Appendix 1

Main events in the history of superconductivity before the discovery by Müller and Bednorz

Year	Experimental results	Theoretical advances	Technological applications
1911	Heike Kamerlingh Onnes discovers superconductivity in mercury at 4 K (Leiden)		
1913	Kamerlingh Onnes is awarded the Nobel Prize for Physics for his work on the properties of matter at low temperature		Kamerlingh Onnes expresses a fairly clear vision of the technical possibility of building high-field magnets using superconducting materials
1933	W. Meissner and R. Ochsenfeld discover the expulsion of the magnetic field when a superconductor is cooled below its critical temperature		
mid-1930s		Several theorists propose phenomenological theories for superconductivity	
1946			Commercial version of helium liquifier developed at MIT
1948	Bardeen, Shockley, and Brittain discover the transistor		

1950	Discovery of the "isotope effect"	Landau and Ginzburg formulate a new phenomenological theory of superconductivity. They are the first to state that there are two types of superconductors (I and II), showing resistivity at different critical fields. Thus only materials of type II are useful for applications.
1950/52	B. Matthias and J. Hulm start systematic work on classifying all known superconducting materials. Two empirical rules emerge: 1 – The average number of electrons per atom should be about 4.55 in the case of niobium, vanadium, and zirconium alloys. 2 – The structure of the material should be cubic. Hulm and Hardy discover superconductivity in V_3Si at 17 K	
1955		Building of the first successful superconducting magnet (0.7 Tesla)
1957	J. Bardeen, L. Cooper, and R. Schrieffer propose a theory explaining all phenomena in superconductivity known so far (BCS theory)	
1958	E. Kunzler discovers the Nb_3Sn is superconducting even in a magnetic field as high as 8.8 Tesla	On the basis of Kunzler's finding it becomes possible to construct big superconducting magnets. The first to do are laboratories for high-energy physics (Argonne, Fermilab, CERN), for use with their bubble chambers.

Appendix 1 (cont.)

Year	Experimental results	Theoretical advances	Technological applications
1962		Brian Josephson predicts pair tunneling between weakly linked superconductors	Scientists at Westinghouse develop the first commercial superconducting niobium-titanium wire
1963	The first so-called "Josephson junction" is produced, verifying the prediction that Copper pairs can tunnel across a barrier as easily as single electrons		Researchers at Bell Labs. reach 7.0 Tesla with a Nb_3Sn superconducting high-field magnet
1964		W. A. Little predicts the existence of organic superconductors with every high transition temperatures	SQUIDS (Superconducting Quantum Interference Devices) are first investigated as very sensitive devices for measuring small magnetic fields
1968			IBM starts a program to develop an entire computer using Josephson junction technology
1972		Bardeen, Cooper, and Schrieffer are awarded the Nobel Prize for their superconductivity theory	
1973	Scientists at Westinghouse reach superconductivity at 23 K by depositing Nb_3Ge on a substrate using thin-film technology; this is the highest temperature reached so far		
1973/75	Discovery of the first oxide superconductors; however, their critical temperatures are about 13 K		

1974   Brian Josephson and Ivar Giaever
        are awarded the Nobel Prize for
        their work on the tunneling effect
        in connection with superconductors

1980   Jerome et al. discover first organic
        superconductors; however, their T_c is
        extremely low

1980s   Numerous technological
        appliactions:
        – magnetic resonance imaging
          devices
        – radiofrequency devices
        – R&D magnets (fusion research,
          high-energy physics, etc.)
        – prototype energy storage

Appendix 2

Chronology of important events in high-temperature superconductivity in the early phase[a]

Year	Month	Event
1986	January	K. A. Müller and G. Bednorz at IBM Zurich discover a ceramic compound of lanthanum, barium, copper, and oxygen that becomes superconducting at 35 K
	April	Article submitted to *Zeitschrift für Physik*
	September	Article appears in *Zeitschrift für Physik*. Tanaka and Chu report on the reproduction of M&B's results
	December	Fall meeting of the Materials Research Society. Chu reaches 40 K in BaLaCuO superconductors at high pressure. Cava discovers LaSrCuO, which becomes superconducting at 36 K
1987	January	P. Chu of the University of Houston and M.-K. Wu of the University of Alabama find a new material superconducting at 98 K by substituting yttrium for lanthanum (YBaCuO); this renders liquid nitrogen cooling adequate
	March	Meeting of the American Physical Society "Woodstock of Physics"
	July	President Reagan announces the 11-point superconductivity initiative and calls for a national conference on future applications of these new materials
	October	Müller and Bednorz are awarded the Nobel Prize for their recent discovery
	December	Japanese researchers (H. Maeda *et al.*) discover a BiCaSrCuO superconductor with a critical temperature of 110 K. (A new high-T_c oxide superconductor with no rare earth element.)
1988	January	Z. Z. Sheng and A. Hermann of the University of Arkansas produce a new superconductor containing calcium and thallium with critical temperature of 120–125 K
	March	Cava *et al.* discover the first HTS material not containing copper, BaKBiO, which becomes superconducting at 30 K. This is the highest value reached so far in non-copper containing compounds

[a]For an extensive glossary and details on applications as well as theoretical and experimental advances in superconductivity see Hunt (1989).

References

Abelson, P. H. 1988 'Federal policies in transition.' *Science* **242**: 4886

Anderson, C. 1991a 'Teething troubles for UK technology labs.' *Nature* **352**: 270–1

Anderson, C. 1991b 'US plans to rescue materials research.' *Nature* **350**: 365

Anderson, Ph. W. & Schrieffer, R. 1991 'A dialogue on the theory of high T_C superconductivity.' *Physics Today* **6**: 54–61

Arthur, W. B. 1988 'Competing technologies: an overview', in G. Dosi *et al.* (eds.) *Technical Change and Economic Theory* (London: Pinter Publishers): 590–607

Ausubel, J. H. & Marchetti, C. 1993. 'Elektron', paper prepared for The Rockefeller University Program for the Human Environment. Manuscript.

Becher, T. 1989 *Academic Tribes and Territories. Intellectual Inquiry and the Cultures of Disciplines* (Milton Keynes: Open University Press)

Bednorz, G. J. & Müller, K. A. 1986 'Possible high T_c superconductivity in the Ba–La–Cu–O system.' *Zeitschrift für Physik B* **64**: 189–93

Bednorz, G. J. & Müller, K. A. 1987 *Perovskite-type Oxides – The New Approach to High Tc Superconductivity* (Stockholm: Nobel Lecture): 63–98

Bednorz, G. J., Takashige, M. & Müller, K. A. 1987a 'Susceptibility measurements support high T_c superconductivity in the Ba–La–Cu–O system.' *Europhysics Letters* **3**: 379

Bednorz, G. J., Müller, K A. & Takashige, M. 1987b 'Superconductivity in alkaline earth-substituted La_2CuO_{4-y}.' *Science* **236**: 73–5

Bijker, W. E., Hughes, T. P. & Pinch, T. J. (eds.) 1989 *The Social Construction of Technological Systems* (Cambridge, Mass: MIT Press)

Binning, G. 1989 *Aus dem Nichts. Über die Kreativität von Natur und Mensch* (München: Piper)

BMFT. 1990 *Faktenbericht 1990 zum Bundesbericht Forschung 1988*. (Bonn)

Boden, M. A. 1994 *Dimensions of Creativity* (Cambridge, Mass.: MIT Press)

Brooks, H. 1987 'National rivalries and international science and technology', in K. Vak (ed.) *Complexities of the Human Environment: A Cultural and Technological Perspective*. (Vienna: Europa-Verlag): 49–62

Bush, V. 1945 *Science – The Endless Frontier. A report to the President on a Program for Postwar Scientific Research*, July 1945; reprinted July 1960 (Washington, D.C.: National Science Foundation)

Callon, M. (ed.) 1989 *La Science et ses Réseaux. Genèse et Circulation des Faits Scientifiques* (Paris: Editions de la Découverte)

Cambrosio, A., Limoges, C. & Pronovost, D. 1990 'Representing biotechnology: An ethnography of Quebec science policy.' *Social Studies of Science* **2**: 195–227

Campbell, D. F. J. 1993 'Strukturen und Modelle der F&E-Finanzierung in Deutschland: eine Policy-Analyse.' *IHS Reihe Politikwissenschaft*, 9

Cava, R. J. 1993 'Superconductors and baseballs.' *Nature* **364**: 297

Cava, R. J. *et al.* 1987 'Bulk superconductivity at 36 K in $La_{1.8}Sr_{0.2}CuO_4$.' *Physical Review Letters* **58**: 408–10

Cava, R. J. *et al.* 1988 'Superconductivity near 30 K without copper: the $Ba_{0.6}K_{0.4}BiO_3$ perovskite.' *Nature* **332**: 814–16.

Chu, C. W. *et al.* 1987 'Evidence for superconductivity above 40 K in the La–Ba–Cu–O compound system.' *Physical Review Letters* **58**: 405–7

Cloître, M. & Shinn, T. 1986 'Enclavement et diffusion du savoir.' *Informations sur les Sciences Sociales* **25**(1): 161–87

Cohendet, P., Ledoux, M. J. & Zuscovitch, E. 1987 *Les Materiaux Nouveaux: Commission des Communautés Européennes* (Paris: Economica)

Cozzens, S. E. 1990 'Options for the future of research evaluation', in Cozzens *et al.* (eds.): 281–94

Cozzens, S. E. *et al.* (eds.) 1990 *The Research System in Transition* (Dordrecht: Kluwer Academic Publishers)

Crawford, E. *et al.* (eds.) 1992 *Denationalizing Science: The Contexts of International Scientific Practice.* Sociology of the Sciences Yearbook (Dordrecht: Kluwer Academic Publishers)

Crow, M. 1989a *New Technology Planning and Development in the United States and Japan* (Ames)

Crow, M. 1989b 'Technology development in Japan and the United States: lessons from the high-temperature superconductivity race.' *Science and Public Policy* **16**(6): 322–44

Dahl, P. F. 1992 *Superconductivity: Its Historical Roots and Development from Mercury to the Ceramic Oxides* (New York: American Institute of Physics)

David, P. A. 1985 'Clio and the economics of QWERTY.' *American Economic Review* **75**: 322–37

David P. A., Mowery, D. & Steinmuller, W. 1988 *The Economic Analysis of Payoffs from Basic Research – An Examination of the Case of Particle Physics Research* (California: Center for Economic Policy Research, Stanford University)

DeBresson, Ch. 1995 'Predicting the most likely diffusion sequence of a new technology through the economy: the case of superconductivity.' *Research Policy* **24**: 685–705

Derian, J. C. 1990 *America's Struggle for Leadership in Technology* (Cambridge, Mass: MIT Press)

Dertouzos, M. L., Lester, R. K., Solow, R. & the MIT Commission on Industrial Productivity. 1989 *Made in America: Regaining the Productive Edge* (Cambridge, Mass.: MIT Press)

de Solla Price, D. 1983 *Sealing Wax and String: A Philosophy of the Experimenter's Craft and its Role in the Genesis of High Technology.* Sarton Lecture, American Association for the Advancement of Science

Di Salvo, F. J. 1987 'New and artificial structured electronic and magnetic materials', in Psaras, P. A. & Langford, H. D. (eds.): 161–76

Dickson, D. 1984 *The New Politics of Science* (Chicago: University of Chicago Press)

Dosi, G. 1988 'Sources, procedures and microeconomic effects of innovation.' *Journal of Economic Literature* **26**: 1120–71

Edge, D. 1990 'Competition in modern science', in T. Frängsmyr (ed.) *Solomon's House Revisited. The Organization and Institutionalization of Science.* Nobel Symposium 75 (Canton, M.A.: Science History Publications): 208–32

Eldridge, J. 1993 *Getting the Message – News, Truth and Power* (London: Routledge)

Etzkowitz, H. 1990 'The second academic revolution: The role of the research university in economic development', in Cozzens *et al.* (eds.): 109–24

Fayard, P. 1993 *Sciences aux Quotidiens* (Nice: Z'Editions)

Felderer, B. & Campbell, D. F. J. 1994 *Forschungsfinanzierung in Europa. Trends – Modelle – Empfehlungen für Österreich* (Wien: Manz)

Felt, U. 1992 'Competition in science and the media', Manuscript of a talk given at the 4S/EASST Conference, Gothenburg 12–15 August 1992

Felt, U. 1993 'Fabricating scientific success stories.' *Public Understanding of Science* 2: 375–90

Felt, U. 1996. 'Conquering a new research territory: University researchers and the discovery of high-temperature superconductivity'. Manuscript submitted to *Social Studies of Science.*

Felt, U. & Nowotny, H. 1992 'Striking gold in the 1990s: The discovery of high-temperature superconductivity and its impact on the science system'. *Science, Technology and Human Values* 17: 506–31

Felt, U. & Nowotny, H. (eds.) 1993 'Science meets the public – a new look at an old problem.' *Public Understanding of Science* 2: 285–426

FOM. 1989 *Report of the Governmental Mission on High Temperature Superconductors of the Netherlands and Japan.* (Utrecht: FOM Office)

Foner, S. & Schwartz, B. (eds.) 1974 *Superconducting Machines and Devices; Large Scale Applications.* Nato Advanced Study Institute (New York: Plenum Press)

Franklin, M. N. 1988 *The Community of Science in Europe. Preconditions for Research Effectiveness in European Community Countries* (Aldershot: Gower)

Friedman, S., Dunwoody, S. & Rogers, C. 1986 *Scientists and Journalists. Reporting Science as News* (New York: The Free Press)

Gardner, H. 1993 *The Creators of the Modern Era* (New York: Basic Books)

Gavroglu, K. & Goudaroulis, O. 1989 *Methodological Aspects of the Development of Low Temperature Physics, 1881–1956: Concepts out of Context(s)* (Dortrecht: Kluwer Academic Publishers)

Geballe, T. H. & Hulm, J. K. 1988 'Superconductivity – the state that came in from the cold.' *Science* 239: 367–75

Geiger, R. 1993 *Research and Relevant Knowledge: American Research Universities since World War II* (New York and Oxford: Oxford University Press)

Gibbons, M., Limoges, C., Nowotny, H., Schwartzman, S., Scott, P. & Trow, M. 1994 *The New Production of Knowledge – the Dynamics of Science and Research in Contemporary* Societies (London: SAGE)

Gilbert, G. N. 1977 'Competition, differentiation and careers in science.' *Social Science Information* 16: 103–23

Goodwin, I. 1991 'Depressed by lack of grants and jobs, materials scientists warm to OSTP plan.' *Physics Today* 6: 89–92

Hacking, I. 1992 'The self-vindication of the laboratory sciences', in A. Pickering (ed.) *Science as Practice and Culture* (Chicago: University of Chicago Press): 29–64

Hazen, R. 1988 *The Breakthrough: The Race for the Superconductor* (New York: Summit)

Heppenheimer, T. A. 1987 *Superconductivity: Research, Applications, and Potential Markets.* (Pasha)

Hilgartner, S. 1990 'The dominant view of popularization: conceptual problems, political issues.' *Social Studies of Science* 20: 519–39

Holton, G. 1986 *The Advancement of Science, and its Burdens* (Cambridge: Cambridge University Press)

Hughes, T. P. 1983 *Networks of Power– Electrification in Western Society*, 1880–1930 (Baltimore: Hopkins)

Hughes, T. P. 1989 *American Genesis. A Century of Invention and Technological Enthusiasm* (New York: Knopf)

Hunt, V. D. 1989 *Superconductivity Sourcebook* (New York: John Wiley & Sons)

Irvine, J., Martin, B. & Isard, Ph. 1990 *Investing in the Future: An International Comparison of Government Funding of Academic and Related Research* (Aldershot: Edward Elgar)

Isawa, Y. 1974 'High speed magnetically levitated and propelled mass ground transportation', in S. Foner & B. Schwartz (eds.) 1974: 347–399

Jacob, F. 1981 *Le Jeu des Possibles. Essai sur la Diversité du Vivant* (Paris: Fayard)

Jacobi D. & Schiele, B. (eds.) 1988 *Vulgariser la Science – Le Procès de l'Ignorance* (Seyssel: Editions Champs Vallon)

Jacques, J. 1990 *L'Imprévu ou la Science des Objets Trouvés* (Paris: Editions Odile Jacob)

Jansen, D. 1991 *Die Supraleitungsforschung und -fërderung in der Bundesrepublik Deutschland nach der Entdeckung der Hochtemperatursupraleitung*, Forschungsbericht (Köln: Max Planck Institut für Gesellschaftsforschung)

Jansen, D. 1994 'National research systems and change: The reaction of the British and German research system to the discovery of high-T_C superconductor.' *Research Policy* **23**: 357–74

Jansen, D. 1995 'Convergence of basic and applied research? Research orientation in German high-temperature superconductor research.' *Science, Technology and Human Values* **20** (2): 197–233

Joerges, B. 1988 'Large technical systems: concepts and issues', in R. Mayntz & T.P. Hughes (eds.) *The Development of Large Technical Systems* (Frankfurt a.M.: Campus): 9–36

Johnson, H. H. 1987 'Introduction. Ensuring contributions to materials science from small-, intermediate-, and large-scale science', in Psaras, P. A. & Langford, H. D. (eds.): 321–3

Kamerlingh Onnes, H. 1967 *Nobel Lecture* 13 December 1913. *Nobel Lectures Physics. Vol 1, 1901–1921* (Amsterdam: Elsevier)

Kevles, D. 1978 *The Physicists: The History of a Scientific Community in Modern America* (New York: Knopf)

Keynan, A. 1991 *The United States as a Partner in Scientific and Technological Cooperation: Some Perspectives from Across the Atlantic* (New York: Carnegie Commission on Science, Technology and Government)

Knorr-Cetina, K. 1984 *Die Fabrikation von Erkenntnis. Zur Anthropologie der Naturwissenschaft* (Frankfurt/Main: Suhrkamp)

Kolata, G. 1987 'Yb or not Yb? That is the question.' *Science* **236**: 663–4

Krohn, W. & Küppers, G. 1989 *Die Selbstorganisation der Wissenschaft* (Frankfurt a.M.: Suhrkamp)

Krohn, W. & Küppers, G. 1990 'Science as a self-organizing system', in Krohn, W., Küppers, G. & Nowotny, H. (eds.): 208–22.

Krohn W. , Küppers, G. & Nowotny, H. (eds.) 1990 *Self-organization. Portrait of a Scientific Revolution. Sociology of the Sciences Yearbook* (Dordrecht: Kluwer Academic Publishers)

Krugman, P. A. 1991 'Myths and realities of U.S. competitiveness.' *Science*, 8 November: 811–15

LaFollette, M. C. 1990 *Making Science our Own – Public Images of Science* 1910–1955 (Chicago: University of Chicago Press)

Latour, B. 1987 *Science in Action. How to Follow Scientists and Engineers Through Society* (Cambridge: Harvard University Press)

Latour, B. & Woolgar, S. 1986 *Laboratory Life: The (Social) Construction of Scientific Facts* (Beverly Hills, C.A.: SAGE)

Lederman, L. 1991 *Science – The End of the Frontier*. Report to the Board of Directors of the AAAS

Lewenstein, B. 1995a 'From Fax to facts: Communication in the cold fusion saga.' *Social Studies of Science* **25**: 403–36

Lewenstein, B. 1995b 'Do public electronic boards help create scientific knowledge? The cold fusion case'. *Science, Technology, and Human Values* **20**: 123–49

Limoges, C. 1993 'Expert knowledge and decision-making in controversy contexts.' *Public Understanding of Science* **2**: 417–26

MacKenzie, D. 1990 *Inventing Accuracy. A Historical Sociology of Nuclear Missile Guidance* (Cambridge, Mass.: MIT Press)

Maddox, J. 1995 'What future for research in Europe?' *Nature* **377**: 283–4

Mansfield, E. 1991 'Academic research and industrial innovation.' *Research Policy* **20**: 1–12

Maranto, G. 1987 'Superconductivity: Hype vs reality.' *Discover*, August: 23–32

Michel, C., Er-Rakho, L. & Raveau, B. 1985 'The oxygen defect perovskite $BaLa_4Cu_5O_{13.4}$ a metallic conductor.' *Material Research Bulletin* **20**: 667

Montroll, E. W. & Shuler, K. E. 1979 'Dynamics of technological evolution: Random walk model for the research enterprise.' *Proc. Natl. Acad. Sci. USA.* **76**(12): 6030–4

Mowery, D. C. & Rosenberg, N. 1989 *Technology and the Pursuit of Economic Growth* (Cambridge: Cambridge University Press)

Müller, K. A. & Bednorz, J. G. 1987 'The discovery of a class of high-temperature superconductors.' *Science*, 4 September: 1133–9

National Advisory Committee on Semiconductors 1989 *A Strategic Industry at Risk* (Arlington, V.A. November 1989)

Nelkin, D. 1987 *Selling Science: How the Press Covers Science and Technology* (New York: Freeman)

Nöldechen, A. 1988 *Die Supraleitung. Nobelpreis für eine technische Revolution* (Düsseldorf: Econ Verlag)

Nowotny, H. 1990 'Individual autonomy and autonomy of science: The place of the individual in the research system', in Cozzens *et al.* (eds.): 331-44

Nowotny, H. 1995 'The dynamics of innovation. On the multiplicity of the new.' *European Review* **3**(3), July: 209–19

OECD 1987 *Review of National Science and Technology Policies, Austria* (Paris)

Ortoli, S. & Klein, J. 1989 *Histoire et Légendes de la Supraconductivité* (Paris: Calmann-Lévy)

OTA 1988 *Commercializing High-Temperature Superconductivity* (Washington D.C.: Congress of the US)

OTA 1990 *High-Temperature Superconductivity in Perspective* (Washington D.C.: Congress of the US)

Ott, H. R. (ed.) 1992 *Ten Years of Superconductivity: 1980–1990*. (Dordrecht: Kluwer Academic Publishers)

Pavitt, K. 1991 'What makes basic research economically useful?' *Research Policy* **20**: 109–19

Pavitt, K. & Patel, P. 1991 'Technological strategies of the world's largest companies.' *Science and Public Policy.* SPRU 25th anniversary papers, vol.18, no.8, December

Pendlebury, D. 1988a 'On the crest of a superconductivity Tsunami.' *The Scientist*, 16 May: 15–16

Pendlebury, D. 1988b '1987's top research focus: Superconductors.' *The Scientist*, 26 December: 13

Perez, C. 1983 'Structural change and the assimilation of new technologies in social systems.' *Futures* **15**: 357–75

Perez, C. 1989 *Technical Change, Competitive Restructuring and Institutional Reform in Developing Countries.* SPRU Publications, Discussion Paper No.4 (Washington, D.C.: World Bank)

Pippard, A. B. 1977 'The historical context of Josephson's discovery', in B. Schwartz & S. Foner (eds.) *Superconductor Applications: SQUIDs and Machines.* (New York: Plenum Press): 1–20

Pool, R. 1991 'A new role for DARPA.' *Nature*, 19 September

Psaras, P. A. & Langford, H. D. (eds.) 1987 *Advancing Materials Research* (Washington, DC: National Academy Press)

Raz, B. 1987 *High Temperature Superconductivity Research: An Indicator of National Science Policy Responsiveness.* Manuscript

Reich, R. 1991 *The Work of Nations. Preparing Ourselves for 21st Century Capitalism* (New York: Knopf)

Renn, J. 1989 *The Time Scales of Conceptual Traditions in Physics: Galilei and Einstein* (Berlin: Manuscript)

Rheinberger, H.-J. 1994 'All that gives rise to an inscription in general'. Ms. to be published

Rip, A. 1990 'An exercise in foresight: The research system in transition – to what?', in Cozzens *et al.* (eds.): 387–401

Rizzuto, C. 1991 *The Evolution of Physics. From the 'Reductionist' to the 'Complexity' Trend* Manuscript

Robinson, A. L. 1987 'A superconductivity happening.' *Science* **235**: 1571

Root-Bernstein, R. S. 1989 *Discovering* (Cambridge: Harvard University Press)

Rosenberg, N. 1991 'Critical issues in science policy research.' *Science and Public Policy.* SPRU 25th anniversary papers, vol.18, no.6

Rowell, J. M. 1988 'Superconductivity Research: A Different View', *Physics Today* **10**: 33–46.

Roy, R. 1988 'HTSC: Restoring scientific and policy perspective', in C. G. Burnham & R. D. Kane (eds.) *World Congress on Superconductivity. Progress in High Temperature Superconductivity*–Vol. 8 (Singapore, New Jersey, Hong Kong): 27–39

Ruivo, B. 1994 "Phases' or 'paradigms' of science policy?' *Science and Public Policy* **21**(3): 157–67

Salomon, J.-J. 1990 'Terror and scruples', in J.-J. Salomon (ed.) *Science, War, Peace* (Paris: Economica)

Schaffer, S. 1994 'Making up discovery', in M. A. Boden (ed.) *Dimensions of Creativity* (Cambridge, Mass.: MIT Press): 13–51

Schechter, B. 1989 *The Path of No Resistance: the Story of the Revolution in Superconductivity* (New York: Simon and Schuster)

Schubert, A. & Braun, T. 1990 'International collaboration in the sciences 1981–1985.' *Scientometrics* **19**(1–2): 3–10

Schwartz, B. 1987 'Materials research laboratories: reviewing the first twenty-five years', in Psaras, P. A, & Langford, H. D. (eds.): 35–48

Schwartz, B. & Foner, S. (eds.) 1977 *Superconductor Applications: SQUIDs and Machines* (New York and London: Plenum Press)

Seeber, B. 1988 'Hochtemperatursupraleitung: Gegenstandsbereich, Forschungsentwicklung und Anwendungspotential aus der Sicht der Wissenschaften.' *Forschungspolitische Früherkennung* (Bern: Schweizer Wissenschaftsrat)

Senker, J. 1991 'Evaluating the funding of strategic science: Some lessons from British experience.' *Research Policy* **20**: 29–43

Shapin, S. 1990 'Science and the public', in R. C. Olby *et al.* (eds.) *Companion to the History of Modern Science* (London: Routledge): 990–1107

Shapin, S. 1991 'The mind is its own place: Science and solitude in seventeenth-century England.' *Science in Context* **4**(1): 191–218

Shapin, S. 1992 'Why the public ought to understand science-in-the-making.' *Public Understanding of Science* **1**: 27–30

Shapley, D. & Roy, R. 1985 *Lost at the Frontier. US Science and Technology Policy Adrift* (Philadelphia: ISI Press)

Sheng, Z. Z. & Hermann, A. M. 1988 'Bulk superconductivity at 120 K in the Tl–Ca/Ba–Cu–O system.' *Nature* **332**: 138–9

Shinn, T. 1988 'Hiérarchies des chercheurs et formes de recherches.' *Actes de la Recherche*, no.74, septembre: 2–22

Shinn, T. & Whitley, R. 1985 *Expository Science: Forms and Functions of Popularization* (Dordrecht: Reidel)

Simon, R. & Smith, A. 1988 *Superconductors – Conquering Technology's New Frontier* (New York: Plenum Press)

SNF 1987 *Der Schweizerischer Nationalfonds zur Förderung der wissenschaftlichen Forschung Ein Kurzporträt*

Sproull, R. L. 1987 'Materials research laboratories: The early years', in Psaras, P. A. & Langford, H. D. (eds.): 25–34.

Tarascon, J. M. *et al.* 1987 'Superconductivity at 40 K in the oxygen-defect perovskites $La_{2-x}Sr_xCuO_{4-y}$,' *Science* **235**: 1373–6.

Traweek, S. 1988 *Beamtimes and Lifetimes. The World of High Energy Physicists* (Cambridge/Mass: Harvard University Press)

Tuchman, G. 1978 *Making News: A Study in the Construction of Reality* (New York: Basic Books)

Tyszkiewicz, M. Th. 1988 *Emerging British Policy for Superconductivity. The National Committee for Superconductivity and the Interdisciplinary Research Centre for Superconductivity*, Thesis (Brighton: SPRU)

Veyne, P. 1983 *Les Grecs Ont-ils Cru à Leur Mythes?* (Paris: Seuil)

Vidali, G. 1992 *Superconductivity: The Next Revolution?* (Cambridge: Cambridge University Press)

Voss, D. F. 1988 'Superconductivity: The FAX factor.' *Science* **240**: 280–1

Waysand, G. 1987 'Supraconductivité: Evolution propre et rapports involontaires avec la société.' *Esprit* **7**: 40–57

Wu, M. K. *et al.* 1987 'Superconductivity at 93 K in a new mixed-phase Y–Ba–Cu–O compound system at ambient pressure.' *Physical Review Letters* **58**: 908–10

Wynne, B. 1992 'Representing policy constructions and interests in SSK.' *Social Studies of Science* **22**: 575–80

York, H. F. 1987 *Making Weapons, Talking Peace* (New York: Basic Books)

Ziman, J. 1990 'What is happening to science?', in Cozzens *et al.* (eds.): 23–34

Ziman, J. 1994 *Prometheus Bound. Science in a Dynamic Steady State* (Cambridge: Cambridge University Press)

Index

---

[1] In the course of the project this Ministry has been renamed twice: First, Ministry for Science, Research and Art and, since spring 1996, Ministry for Science, Traffic and Art.